T0329573

Autonomous Navigation and Deployment of UAVs for Communication, Surveillance and Delivery

Autonomous Navigation and Deployment of UAVs for Communication, Surveillance and Delivery

Hailong Huang, PhD
Department of Aeronautical and Aviation Engineering
The Hong Kong Polytechnic University, Hong Kong, China

Andrey V. Savkin, PhD
School of Electrical Engineering and Telecommunications
University of New South Wales, Sydney, NSW, Australia

Chao Huang, PhD
Department of Industrial and Systems Engineering
The Hong Kong Polytechnic University, Hong Kong, China

IEEE PRESS
WILEY

Published by John Wiley & Sons, Inc., Hoboken, New Jersey.
Published simultaneously in Canada.

For general information on our other products and services or for technical support, please contact our Customer Care Department within the United States at (800) 762-2974, outside the United States at (317) 572-3993 or fax (317) 572-4002.

Wiley also publishes its books in a variety of electronic formats. Some content that appears in print may not be available in electronic formats. For more information about Wiley products, visit our web site at www.wiley.com.

Library of Congress Cataloging-in-Publication Data Applied for:
Hardback ISBN: 9781119870838

Cover Design: Wiley
Cover Image: © yuanyuan yan/Getty Images

Set in 9.5/12.5pt STIXTwoText by Straive, Chennai, India

Contents

Author Biographies

Hailong Huang received the BSc degree in automation, from China University of Petroleum, Beijing, China, in 2012, and received the PhD degree in Systems and Control from the University of New South Wales, Sydney, Australia, in 2018. He is an Assistant Professor at the Department of Aeronautical and Aviation Engineering at The Hong Kong Polytechnic University, Hong Kong. His current research interests include guidance, navigation, and control of mobile robots, multi-agent systems, and distributed control.

Andrey V. Savkin received the MS and PhD degrees in mathematics from the Leningrad State University, Saint Petersburg, Russia, in 1987 and 1991, respectively. From 1987 to 1992, he was with the Television Research Institute, Leningrad, Russia. From 1992 to 1994, he held a Postdoctoral position in the Department of Electrical Engineering, Australian Defence Force Academy, Canberra. From 1994 to 1996, he was a Research Fellow in the Department of Electrical and Electronic Engineering and the Cooperative Research Centre for Sensor Signal and Information Processing, University of Melbourne, Australia. From 1996 to 2000, he was a Senior Lecturer, and then an Associate Professor in the Department of Electrical and Electronic Engineering, University of Western Australia, Perth. Since 2000, he has been a Professor in the School of Electrical Engineering and Telecommunications, University of New South Wales, Sydney, NSW, Australia. His current research interests include robust control and state estimation, hybrid dynamical systems, guidance, navigation and control of mobile robots, applications of control and signal processing in biomedical engineering and medicine. He has authored/coauthored seven research monographs and numerous journal and conference papers on these topics. He has served as an Associate Editor for several international journals.

Chao Huang received the BSc degree in automation, from China University of Petroleum, Beijing, China, in June 2012, and received the PhD degree from the University of Wollongong, Wollongong, Australia, in December 2018. She is a research assistant professor at the Department of Industrial and Systems Engineering, the Hong Kong Polytechnic University, Hong Kong. Her interests include motion planning, human machine collaboration, fault tolerant, automotive control and application.

Preface

Unmanned aerial vehicles (UAVs), also known as aerial drones, have started to reshape our modern life, thanks to the inherent attributes such as mobility and flexibility. Once national legislations allow UAVs to fly autonomously, swarms of UAVs will populate our city skies to conduct various missions: rescue operations, surveillance, and monitoring, and also some emerging applications such as goods delivery and telecommunications.

This book is primarily a research monograph that presents, in a detailed and unified manner, the recent advancements relevant to the applications of UAVs in wireless communications, surveillance and monitoring of ground targets and areas, and goods delivery. The main intended audience for this monograph includes postgraduate and graduate students, as well as professional researchers and industry practitioners working in a variety of areas such as robotics, aerospace engineering, wireless communications, signal processing, system theory, computer science and applied mathematics who have an interest in the growing field of autonomous navigation and deployment of UAVs. This book is essentially self-contained. The reader is assumed to be familiar with basic undergraduate level mathematical techniques. The results presented are discussed to a great extent and illustrated by examples. We hope that readers find this monograph interesting and useful and gain a deeper insight into the challenging issues in the field of autonomous navigation and deployment of UAVs for communication, surveillance, and delivery. Moreover, in the book, we have made comments on some open issues, and we encourage readers to explore them further. The material in this book derives from a period of research collaboration between the authors from 2018 to 2022. Some of its parts have separately appeared in journal and conference papers. The manuscript integrates them into a unified whole, highlights connections between them, supplements them with new original findings of the authors, and presents the entire material in a systematic and coherent fashion.

In preparation of this research monograph, the authors wish to acknowledge the financial support they have received from the Australian Research Council.

This research work has also received funding from the Australian Government, via grant AUSMURIB000001 associated with ONR MURI grant N00014-19-1-2571. Also, the authors are grateful for the support they have received throughout the production of this book from the School of Electrical Engineering and Telecommunications at the University of New South Wales, Sydney, Australia, the Department of Aeronautical and Aviation Engineering and the Department of Industrial and Systems Engineering, The Hong Kong Polytechnic University, Hong Kong, China.

Furthermore, Andrey Savkin is grateful for the love and support he has received from his family. Hailong Huang and Chao Huang are also grateful for the support from their parents.

Hailong Huang
Andrey V. Savkin
Chao Huang

1

Introduction

1.1 Applications of UAVs

Thanks to the inherent attributes such as mobility and flexibility, unmanned aerial vehicles (UAVs), also known as aerial drones, have started to reshape our modern life. Once national legislations make laws to allow UAVs to fly autonomously, swarms of UAV will populate the sky of our cities to conduct various missions: rescue operations, surveillance, and monitoring, and also some emerging applications such as goods delivery and telecommunications. Some typical applications of UAVs are summarized in Figure 1.1.

UAVs can service humans. A typical example is that UAVs play the role of aerial base stations to provide communication service to cellular users, especially in some congested urban areas [1]. This is a promising solution to 5G and beyond-5G networks. It is also very useful in disaster areas where the communication infrastructures are down. Also, UAVs have been used to track targets, such as humans, animals, and vehicles [2], and in agriculture [3], traffic monitoring [4], architecture inspection [5], environment monitoring [6], disaster management [7]. Furthermore, UAVs can provide service to wireless sensor networks (WSNs) [8]. Working as the aerial sinks, the UAVs can collect sensory data from distributed sensor nodes. They can navigate ground robots since they may have a better view of the environment, and they can also collaborate with ground robots to execute complex tasks. Beyond those presented in Figure 1.1, package delivery is another service UAVs can provide, which also attracts great interest from both the research community [9] and logistics companies [10–14].

In addition to the sole-UAV usage, the collaboration between UAVs and ground vehicles for surveillance and parcel delivery has gained attention. A key point is using one or more UAVs to visit a given set of positions. Considering the limited capacity of the on-board battery, the flying time of a UAV is constrained. A straightforward idea is to install ground charging stations at which a UAV can recharge or replace its battery [15]. Where to deploy the ground charging stations influences

Autonomous Navigation and Deployment of UAVs for Communication, Surveillance and Delivery, First Edition. Hailong Huang, Andrey V. Savkin, and Chao Huang.
© 2023 The Institute of Electrical and Electronics Engineers, Inc. Published 2023 by John Wiley & Sons, Inc.

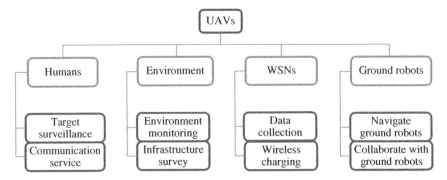

Figure 1.1 The usage of UAVs.

the coverage performance that UAVs can achieve. Given a set of ground charging stations, the review [16] focuses on the path planning problem of a UAV to visit a given set of positions successfully. Another idea is to use ground robots to function as mobile charging platforms [17–19]. Specifically, the chapter [17] considers the usage of a battery-constrained UAV and a battery-unlimited ground robot for large-scale mapping. The UAV can recharge its battery on the ground robot. While the ground robot can only move on the road network, the UAV can traverse the areas off the road network. The authors provide a strategy for the cooperation of the UAV and ground robot such that they can finish the mapping mission under the energy constraint of the UAV. The chapter [18] considers a similar scenario as [17]. The authors provide an integer program for this problem. Different from [17, 18], in [19], the UAV can travel with a ground robot together and recharge its battery during the movement. Clearly, this strategy reduces the time to finish the mission.

1.2 Problems of Autonomous Navigation and Deployment of UAVs

In order to fully reap the benefits of UAVs in the aforementioned real-life applications, some core technical challenges, including the 3D placement of multiple UAVs, the trajectory/movement design, the energy efficiency optimization. These deployment and autonomous navigation of UAVs play an extremely important role for the usage of UAVs. A typical scenario is that a team of collaborating UAVs is conducting a mission for which it is needed to determine some optimal operation status including physical positions and other application-dependent attributes such as transmission powers when UAVs serve as aerial base stations.

For a static situation, we can formulate some UAV navigation and deployment problems as optimization problems. The objective function can be application-dependent. For example, when UAVs are used to serve cellular users, a typical problem is how to deploy UAVs to cater to wireless users' instantaneous traffic demands. Existing research has investigated the trajectory planning problem for a single UAV to relay information [20] and broadcast/multicast data packets [21]. Besides the trajectory planning problem, researchers have also investigated the UAV deployment so that wireless coverage is provided to the static users in a target region, by designing the optimal operating location in 3D space [22], minimizing the number of the stop points for the UAV [23], and minimizing the total deployment time [24]. When UAVs are used to monitor ground targets, some metric describing the quality of surveillance will be regarded as the objective. Moreover, the deployment problem often comes with some constraints. An important constraint is the connectivity [25]. When multiple UAVs operate together, they need to form a connected network with some ground base station for communication. The common approach is to introduce a connectivity graph, which can restrict the relative positions of the UAVs so that a valid communication channel between a pair of UAVs is guaranteed. Another constraint is about collision avoidance. For a particular area, there may be existing some infrastructures such as buildings, which may be regarded as no-fly zones [26]. Such no-fly zones further place some constraints to the UAV deployment problem. Then, the deployment problem becomes a constrained optimization problem, and the solution to this problem is the positions of the UAVs.

For a dynamic situation, the optimal positions of UAVs will be time-varying. In this case, deployment and navigation of UAVs are coupled. The optimal positions of UAVs are computed by addressing the deployment problem, and then the UAVs are navigated from their current positions to new positions. During the navigation process, it should be guaranteed that the connectivity is maintained, UAVs do not collide with any obstacles and do not enter any no-fly zones. Model predictive control (MPC) [27] has been recognized as an important tool to address this type of constrained optimization problems. A review of recent results on deployment and navigation of teams of collaborating UAVs for surveillance can be found in the survey paper [28]. Moreover, a review of challenges and achievements in reaching full autonomy of UAVs is presented in [29]. The research monograph [30] studies various applications of UAVs for support of wireless communication networks.

Though the research community has already made a great contribution to the navigation and deployment of UAVs, many of the existing approaches suffer from the complexity for real-time implementation. Additionally, the mobility of ground targets (in wireless coverage and also surveillance applications) is generally overlooked by many research articles, based on which, the mobility of UAVs needs to be carefully considered to get a better quality of service (QoS). Motivated by such research gaps, and to facilitate the application of UAVs, navigation and deployment methods should be implemented in real-time at each UAV using local information only. This requires proposed methods be computationally efficient. Moreover, the optimality of the overall performance of the UAVs should be guaranteed. Therefore, decentralized algorithms are often needed for UAV deployment and navigation.

1.3   Overview and Organization of the Book

In this section, we briefly describe the results presented in this research monograph.

This book is problem-oriented, not technique-oriented. So each chapter is self-contained and devotes to a detailed discussion of an interesting problem that arises in the rapidly developing area of UAVs' applications. We present relevant approaches from a control system viewpoint. Thus, in Chapters 2–6, we first present system models and then formulate problems of interest, which are followed by proposed approaches to address the problems. Finally, we present computer simulation results to illustrate the effectiveness of the proposed approaches. The organization of the book is as follows.

In Chapter 2, we discuss an application of UAVs in providing cellular service as aerial base stations. We study a problem of proactive UAV deployment. The deployment of UAVs plays a key role for the quality of service in such applications. Two typical scenarios are studied. The first scenario is in urban areas, and the UAVs are deployed over streets to avoid collision with buildings. The second scenario is for disaster areas. We formulate several optimization problems to optimize the quality of service provided by the UAVs, and computationally efficient algorithms are presented to address these problems.

Chapter 3 discusses some recent developments in using UAVs to monitor ground areas and targets. Specifically, we present approaches to finding the minimum number of UAVs equipped with ground-facing video cameras and their deployment positions to fully monitor an area of interest, which can be either a flat area or an uneven area with buildings, hills, or mountains. We also present algorithms that can find the optimal positions of UAVs to survey a group of ground targets within a certain area. We develop deployment algorithms for both the 2D and 3D deployment of UAVs. Theoretical analysis on the performance of these approaches is also provided.

In Chapter 4, we discuss applications of UAVs for surveillance and monitoring of ground areas and targets, which corresponds to various practical applications including but not limited to surveillance of disaster processes such as offshore oil spills, flood and coal ash spills, and monitoring ground vehicles and pedestrians.

We present several decentralized algorithms for navigation of a team of UAVs to collaboratively conduct surveillance missions. The properties of these algorithms such as optimality are discussed.

Chapter 5 focuses on covert video surveillance using UAVs, which is a relatively new research area. Different from usual surveillance applications discussed in Chapters 3 and 4, covert surveillance requires that the intention of the UAVs is not discovered by the targets of interest. We present two approaches to this problem. The first approach is optimization-based. We present a new metric to characterize the disguising performance, which evaluates the change of the relative distance and angle between the UAV and the target. Then, we formulate an optimization problem, which jointly maximizes the disguising performance and minimizes the energy efficiency of the UAV, subject to the motion constraint of the UAV and the requirement of keeping the target within view. We present a dynamic programming method to plan the UAV's trajectory in an online manner. The second approach is a biologically inspired motion camouflage-based method. To achieve motion camouflage, the UAV always moves on the straight line segment connecting the target and a fixed reference point. A sliding mode control strategy is developed, which only takes the bearing information as input. We present extensive computer simulations to demonstrate the performance of these approaches.

In Chapter 6, we discuss the applications of UAVs in the last-mile parcel delivery. UAVs have been considered as a promising tool for future logistics industry by many companies thanks to reduced cost and increased mobility. However, one barrier is the limited flight time due to the limitation of onboard batteries. This chapter presents recent research results on using public transportation vehicles to assist UAV delivery. A particular attention is paid to path planning problems when UAVs can travel with public transportation vehicles, and several algorithms are presented to deal with these problem in different situations.

1.4 Some Other Remarks

Chapter 2 of this book studies using UAVs for wireless communication coverage, Chapters 3–5 are about using UAVs for video surveillance of ground areas and targets, and Chapter 6 concentrates on UAV assisted delivery. On the other hand, Chapters 3 and 3 of this book studies UAV deployment, Chapters 4 and 5 address UAV navigation, and Chapter 6 concentrates on UAV flight scheduling. Furthermore, Chapters 1–4 studies teams of UAVs, Chapter 5 concentrates on using a single UAV, whereas Chapter 6 studies UAVs collaborating with ground public transportation vehicles. Also, it should be pointed out that Chapters 2–5 study deterministic models that often contain large uncertainties, whereas Chapter 6 addresses both deterministic and stochastic models.

It should be pointed out that teams of collaborating autonomous UAVs guided by decentralized navigation algorithms developed in this book can be naturally viewed as networked control systems; see e.g. [31] and references therein.

The main results of this research monograph were originally published in the journal papers [32–49].

The literature in the field of autonomous UAV navigation and deployment for communication, ground surveillance, and parcel delivery is vast, and we have limited ourselves to references that we found most useful or that contain material supplementing this text. The coverage of the literature in this book is by no means complete. We apologize in advance to many authors whose valuable research contributions have not been mentioned.

In conclusion, the area of autonomous navigation and deployment of UAVs is a fascinating discipline bridging robotics, aerospace engineering, system theory, control engineering, communications, information theory, computer science, and applied mathematics. The study of decentralized UAV navigation and deployment problems represents a difficult and exciting challenge in system engineering. We hope that this research monograph will help in some small way to meet this challenge.

References

1 H. Huang and A. V. Savkin, "Towards the internet of flying robots: a survey," *Sensors*, vol. 18, no. 11, p. 4038, 2018.

2 L. D. P. Pugliese, F. Guerriero, D. Zorbas, and T. Razafindralambo, "Modelling the mobile target covering problem using flying drones," *Optimization Letters*, vol. 10, no. 5, pp. 1021–1052, 2016.

3 C. Zhang and J. M. Kovacs, "The application of small unmanned aerial systems for precision agriculture: a review," *Precision Agriculture*, vol. 13, no. 6, pp. 693–712, 2012.

4 K. Kanistras, G. Martins, M. J. Rutherford, and K. P. Valavanis, "A survey of unmanned aerial vehicles (UAVs) for traffic monitoring," in *Handbook of Unmanned Aerial Vehicles*, (eds. K.P. Valavanis, G.J. Vachtsevanos), pp. 2643–2666, Springer, 2014.

5 Z. Zhou, C. Zhang, C. Xu, F. Xiong, Y. Zhang, and T. Umer, "Energy-efficient industrial internet of UAVs for power line inspection in smart grid," *IEEE Transactions on Industrial Informatics*, vol. 14, no. 6, pp. 2705–2714, 2018.

6 B. Esakki, S. Ganesan, S. Mathiyazhagan, K. Ramasubramanian, B. Gnanasekaran, B. Son, S. W. Park, and J. S. Choi, "Design of amphibious vehicle for unmanned mission in water quality monitoring using internet of things," *Sensors*, vol. 18, no. 10, p. 3318, 2018.

7 S. M. Adams and C. J. Friedland, "A survey of unmanned aerial vehicle (UAV) usage for imagery collection in disaster research and management," in *The 9th International Workshop on Remote Sensing for Disaster Response*, p. 8, 2011.

8 C. Caillouet and T. Razafindralambo, "Efficient deployment of connected unmanned aerial vehicles for optimal target coverage," in *Global Information Infrastructure and Networking Symposium (GIIS)*, pp. 1–8, IEEE, 2017.

9 V. Gatteschi, F. Lamberti, G. Paravati, A. Sanna, C. Demartini, A. Lisanti, and G. Venezia, "New frontiers of delivery services using drones: a prototype system exploiting a quadcopter for autonomous drug shipments," in *The 39th Annual Computer Software and Applications Conference (COMPSAC)*, vol. 2, pp. 920–927, IEEE, 2015.

10 Amazon.com Inc., "Amazon prime air," Accessed on 1 Nov. 2021. Online: http://www.amazon.com/primeair.

11 "China is on the fast track to drone deliveries," Accessed on 1 Nov. 2021. Online: https://www.bloomberg.com/news/features/2018-07-03/china-s-on-the-fast-track-to-making-uav-drone-deliveries.

12 "SF express approved to fly drones to deliver goods," Accessed on 1 Nov. 2021. Online: https://www.caixinglobal.com/2018-03-28/sf-express-approved-to-fly-drones-to-deliver-goods-101227325.html.

13 "DHL's parcelcopter: changing shipping forever," Accessed on 1 Nov. 2021. Online: https://discover.dhl.com/business/business-ethics/parcelcopter-drone-technology.

14 "UPS testing drones for use in its package delivery system," Accessed on 1 Nov. 2021. Online: https://www.apnews.com/f34dc40191534203aa5d041c3010f6c5.

15 I. Hong, M. Kuby, and A. T. Murray, "A range-restricted recharging station coverage model for drone delivery service planning," *Transportation Research Part C: Emerging Technologies*, vol. 90, pp. 198–212, 2018.

16 K. Sundar and S. Rathinam, "Algorithms for routing an unmanned aerial vehicle in the presence of refueling depots," *IEEE Transactions on Automation Science and Engineering*, vol. 11, no. 1, pp. 287–294, 2014.

17 P. Maini and P. Sujit, "On cooperation between a fuel constrained UAV and a refueling UGV for large scale mapping applications," in *International Conference on Unmanned Aircraft Systems (ICUAS)*, pp. 1370–1377, IEEE, 2015.

18 Z. Luo, Z. Liu, and J. Shi, "A two-echelon cooperated routing problem for a ground vehicle and its carried unmanned aerial vehicle," *Sensors*, vol. 17, no. 5, p. 1144, 2017.

19 K. Yu, A. K. Budhiraja, and P. Tokekar, "Algorithms for routing of unmanned aerial vehicles with mobile recharging stations and for package delivery," *arXiv preprint, arXiv:1704.00079*, 2017.

20 S. Zhang, H. Zhang, Q. He, K. Bian, and L. Song, "Joint trajectory and power optimization for UAV relay networks," *IEEE Communications Letters*, vol. 22, no. 1, pp. 161–164, 2018.

21 Y. Zeng, X. Xu, and R. Zhang, "Trajectory design for completion time minimization in UAV-enabled multicasting," *IEEE Transactions on Wireless Communications*, vol. 17, no. 4, pp. 2233–2246, 2018.

22 A. Al-Hourani, S. Kandeepan, and S. Lardner, "Optimal LAP altitude for maximum coverage," *IEEE Wireless Communications Letters*, vol. 3, no. 6, pp. 569–572, 2014.

23 M. Mozaffari, W. Saad, M. Bennis, and M. Debbah, "Unmanned aerial vehicle with underlaid device-to-device communications: performance and tradeoffs," *IEEE Transactions on Wireless Communications*, vol. 15, no. 6, pp. 3949–3963, 2016.

24 X. Zhang and L. Duan, "Fast deployment of UAV networks for optimal wireless coverage," *IEEE Transactions on Mobile Computing*, vol. 18, no. 3, pp. 588–601, 2018.

25 Z. Han, A. L. Swindlehurst, and K. R. Liu, "Optimization of MANET connectivity via smart deployment/movement of unmanned air vehicles," *IEEE Transactions on Vehicular Technology*, vol. 58, no. 7, pp. 3533–3546, 2009.

26 H. Huang and A. V. Savkin, "Energy-efficient autonomous navigation of solar-powered UAVs for surveillance of mobile ground targets in urban environments," *Energies*, vol. 13, no. 21, p. 5563, 2020.

27 E. F. Camacho and C. B. Alba, *Model predictive control*. Springer, 2013.

28 X. Li and A. V. Savkin, "Networked unmanned aerial vehicles for surveillance and monitoring: a survey," *Future Internet*, vol. 13, no. 7, p. 174, 2021.

29 T. Elmokadem and A. V. Savkin, "Towards fully autonomous UAVs: a survey," *Sensors*, vol. 21, no. 18, p. 6223, 2021.

30 H. Huang, A. V. Savkin, and C. Huang, *wireless communication networks supported by autonomous UAVs and mobile ground robots*. Elsevier, 2022.

31 A. S. Matveev and A. V. Savkin, *Estimation and control over communication networks*. Boston, MA: Birkhauser, 2009.

32 H. Huang and A. V. Savkin, "An algorithm of efficient proactive placement of autonomous drones for maximum coverage in cellular networks," *IEEE Wireless Communications Letters*, vol. 7, no. 6, pp. 994–997, 2018.

33 H. Huang and A. V. Savkin, "A method for optimized deployment of unmanned aerial vehicles for maximum coverage and minimum interference in cellular networks," *IEEE Transactions on Industrial Informatics*, vol. 15, no. 5, pp. 2638–2647, 2019.

34 A. V. Savkin and H. Huang, "Deployment of unmanned aerial vehicle base stations for optimal quality of coverage," *IEEE Wireless Communications Letters*, vol. 8, no. 1, pp. 321–324, 2019.

35 A. V. Savkin and H. Huang, "Range-based reactive deployment of autonomous drones for optimal coverage in disaster areas," *IEEE Transactions on Systems, Man, and Cybernetics: Systems*, vol. 51, no. 7, pp. 4606–4610, 2021.

36 A. V. Savkin and H. Huang, "Asymptotically optimal deployment of drones for surveillance and monitoring," *Sensors*, vol. 19, no. 9, p. 2068, 2019.

37 A. V. Savkin and H. Huang, "Proactive deployment of aerial drones for coverage over very uneven terrains: a version of the 3D art gallery problem," *Sensors*, vol. 19, no. 6, p. 1438, 2019.

38 A. V. Savkin and H. Huang, "A method for optimized deployment of a network of surveillance aerial drones," *IEEE Systems Journal*, vol. 13, no. 4, pp. 4474–4477, 2019.

39 H. Huang and A. V. Savkin, "An algorithm of reactive collision free 3D deployment of networked unmanned aerial vehicles for surveillance and monitoring," *IEEE Transactions on Industrial Informatics*, vol. 16, no. 1, pp. 132–140, 2020.

40 H. Huang, A. V. Savkin, and C. Huang, "Decentralized autonomous navigation of a UAV network for road traffic monitoring," *IEEE Transactions on Aerospace and Electronic Systems*, vol. 57, no. 4, pp. 2558–2564, 2021.

41 A. V. Savkin and H. Huang, "Navigation of a network of aerial drones for monitoring a frontier of a moving environmental disaster area," *IEEE Systems Journal*, vol. 14, no. 4, pp. 4746–4749, 2020.

42 A. V. Savkin and H. Huang, "Asymptotically optimal path planning for ground surveillance by a team of UAVs," *IEEE Systems Journal*, pp. 1–4, 2021. https://ieeexplore.ieee.org/abstract/document/9580756.

43 A. V. Savkin and H. Huang, "Navigation of a UAV network for optimal surveillance of a group of ground targets moving along a road," *IEEE Transactions on Intelligent Transportation Systems*, pp. 1–5, 2021. https://ieeexplore.ieee.org/document/9430769.

44 H. Huang and A. V. Savkin, "Navigating UAVs for optimal monitoring of groups of moving pedestrians or vehicles," *IEEE Transactions on Vehicular Technology*, vol. 70, no. 4, pp. 3891–3896, 2021.

45 H. Huang, A. V. Savkin, and W. Ni, "Online UAV trajectory planning for covert video surveillance of mobile targets," *IEEE Transactions on Automation Science and Engineering*, vol. 19, no. 2, pp. 735–746, 2022.

46 A. V. Savkin and H. Huang, "Bioinspired bearing only motion camouflage UAV guidance for covert video surveillance of a moving target," *IEEE Systems Journal*, vol. 15, no. 4, pp. 5379–5382, 2021.

47 H. Huang, A. V. Savkin, and C. Huang, "Reliable path planning for drone delivery using a stochastic time-dependent public transportation network," *IEEE Transactions on Intelligent Transportation Systems*, vol. 22, no. 8, pp. 4941–4950, 2021.

48 H. Huang, A. V. Savkin, and C. Huang, "Round trip routing for energy-efficient drone delivery based on a public transportation network," *IEEE Transactions on Transportation Electrification*, vol. 6, no. 3, pp. 1368–1376, 2020.

49 H. Huang, A. V. Savkin, and C. Huang, "Drone routing in a time-dependent network: toward low-cost and large-range parcel delivery," *IEEE Transactions on Industrial Informatics*, vol. 17, no. 2, pp. 1526–1534, 2021.

2

Deployment of UAV Base Stations for Wireless Communication Coverage

2.1 Introduction

Due to the tremendous increase of recent wireless traffic demand, Internet Service Providers (ISPs) have been dedicated to developing effective strategies to improve user experience in cellular networks [1]. Densification of stationary base stations (BSs) is one solution [2]; however, it has various drawbacks such as the high cost of site rental and backhaul links. More importantly, it may not be efficiently utilized in nonpeak periods, which is a waste of precious resource. An alternative solution is to deploy autonomous unmanned aerial vehicles (UAVs), which work as flying BSs, to provide Internet service to user equipments (UEs). Because of the flexibility, the utility of UAVs in cellular networks attracts lots of research on the communication models including the air-to-ground pathloss model [3] and the interference model [4], and the placement problem including the placement of a single UAV [5, 6] and several UAVs [7, 8], etc.

Although the implementation of UAVs in cellular networks requires various techniques to come together, in this chapter, we focus on one key issue: the deployment of UAVs. The positions of UAVs not only influence the coverage of the area of interest but also impact the interference at a certain UE from different UAVs. The interference places challenges on the signal demodulation at UEs, because when the signal to interference and noise ratio is below a threshold, UEs cannot demodulate the intended signal. Therefore, the fundamental question we answer here is how to deploy the UAVs such that they can serve the largest number of UEs and impose the least interference on UEs in a given area.

Different from the assumption in many existing papers, i.e. UEs are randomly scattered or following a predefined distribution (see e.g. [2]), we consider a more practical scenario. In particular, the UEs to be served by UAVs are outdoor. This assumption can make the performance of our approach closer to reality because indoor UEs are not the targets of UAVs. To represent city environments, we adopt a

Autonomous Navigation and Deployment of UAVs for Communication, Surveillance and Delivery,
First Edition. Hailong Huang, Andrey V. Savkin, and Chao Huang.

street graph model. We construct a street graph with a set of points, and each street in the area is represented by a subset of these points. Furthermore, we associate each street point with a UE density function, which reflects the traffic demand at this point during a certain period of time. Such UE density function plays an important role in determining the positions of UAVs. In practice, the UE density function can be constructed via either history recordings or online crowdsensing [9]. In this chapter, we choose the former method, i.e. we build up the UE density functions based on a real dataset collected from a social discovery mobile App: Momo [10], which can be a good reflection of the real UE distribution.

Regarding the feasible positions of UAVs, we have the following concerns. First, to avoid hitting tall buildings, the UAVs are deployed over the streets. Second, the UAVs should be deployed within a certain range of the existing electric power poles (we call it charging pole in the rest of the chapter to indicate its function), where they can recharge the battery.[1] The reason to involve this condition is that the working time of the battery on the state-of-the-art commercial UAVs, such as DJI,[2] is quite limited. Thus, to guarantee on the service time, the flying time between the charging pole and hovering position should be carefully controlled, otherwise, the UAVs may run out of battery. There are some stationary BSs on the street graph and the UAVs together with these BSs that form a connected communication graph. In particular, the BSs work as access points and the UAVs serve UEs. A request from a UE will be sent to a BS by the serving UAV either directly or via other relay UAVs.

We first assume that the UAVs use different frequencies to transmit data to UEs. We formulate an optimization model to maximize the UE coverage and minimize the relay cost between UAVs. We seek the optimal positions on the street graph for the UAVs to optimize the objective subject to that of the UAVs and the BSs to form a connected graph. Second, we formulate an optimization problem seeking the optimal positions of the UAVs on the street graph to maximize UE coverage and minimize the interference effect. This problem is slightly similar to the blanket coverage problem of [11]. However, in the problem considered in this chapter, complete (perfect) blanket coverage is not realistic. Furthermore, unlike [11], we take into account interference. We analyze the properties of local maxima of our problem, based on which, we then propose a distributed algorithm to determine the locally optimal positions for UAVs. In this distributed algorithm, each UAV only requires the local information such as the UE densities and the positions of the nearby UAVs, which makes it superior to some existing centralized algorithms, e.g. [8, 12–14]. We prove that the proposed algorithm converges to the local maxima within a finite number of steps.

---

1 http://www.sbs.com.au/news/article/2017/08/23/powerlinescharge-drones-vic-students.
2 http://www.dji.com.

Another scenario this chapter considers is the use of autonomous UAVs to provide wireless communication services to UEs in a disaster area. After a disaster such as an earthquake, while some people are stuck inside buildings, others are somehow relocated to some open areas such as a public square for a temporary stay. For example, the Sichuan earthquake in 2008 forced about 4.8 million homeless people to relocate [15]. When a tsunami is expected, the government may ask people to leave the area. People may travel by car and follow a safe route, and then a lot of cars will be on the road. Since all or most of the existing cell towers in the affected area are destroyed, or they cannot serve such a high amount of people, people in the area may not get cellular service. In these situations, autonomous UAVs are deployed to service the affected people.

Unlike [12, 13, 16–18] assuming that the UAVs are deployed at the same altitude, we consider the case when the UAVs are deployed at different altitudes to avoid collisions. Following [12, 13], it is assumed that each user is covered by the nearest UAV, i.e. the UAV with the shortest distance to this user. The objective is to minimize the average UAV-user distance. To address this optimization problem, a novel distributed algorithm is proposed. So it has an advantage over centralized algorithms, see e.g. [12, 13]. Instead of assuming the availability of users' locations, the proposed algorithm is based on only the strength of the received signal when the UAV communicates with users and some information that is shared by nearby UAVs. From some initial position, this algorithm reactively navigates each UAV to the center of mass of the set of the users it serves. Unlike the proactive algorithms of [12, 13, 16, 17, 19], the proposed algorithm is reactive and implementable in real-time. Furthermore, it runs on each UAV individually and does not require heavy computations. The convergence of the algorithm is proved, and its performance is evaluated through computer simulations.

The main contributions of this chapter are twofold. First, we formulate some optimization problems for the optimal deployment of UAVs. For the case of urban areas, all the feasible positions of UAVs are at some street points. In the case of disaster areas, they are allowed to move in 3D space. Second, we present decentralized algorithms to navigate the UAVs to at least locally optimal positions. These algorithms are effortless to implement and practical in real applications compared to those centralized ones. The main results of this chapter were originally published in [20–23].

The rest of the chapter is organized as follows. Section 2.2 discusses some closely related work. Section 2.3 presents the deployment method for maximizing coverage. Section 2.4 presents the deployment method for maximizing coverage and minimizing interference. Section 2.5 presents the Voronoi partitioning-based deployment method, and Section 2.6 further presents the range-based reactive deployment method. Finally, Section 2.7 summarizes the chapter.

2.2 Related Work

The utility of UAVs to provide network service in cellular networks is a relatively new research area. With the development of 5G networks, UAVs will play a more and more crucial role as a continuous network support [24]. In this section, we discuss some relevant publications to this chapter and highlight the differences with our work.

One group of publications focuses on the basic wireless communication models, such as the downlink and uplink between UEs and UAVs, and between UAVs and stationary BSs. Here, the wireless communication models are different from the traditional models for ground senders and receivers. For example, in [3], the authors model the air-to-ground link. They show that there are two types of propagation situations: UAVs have line-of-sight (LoS) with UEs and UAVs have nonline-of-sight (NLoS) with UEs due to reflections and diffractions. A closed-form expression is presented for the probability of LoS between UAVs and UEs. The reference [4] focuses on the interference effect. To be specific, the maximum coverage of two UAVs in the presence and absence of interference between them is investigated. Further, the air-to-ground channel models utilizing more information about the environment such as the shapes of the buildings were investigated [25].

Based on these fundamental models, another category considers some interesting applications. For instance, a proactive UAV deployment framework is proposed in [26], which aims at lowering the workload of existing stationary BSs. Consider three typical social activities: stadium, parade, and gathering are considered. Also, a traffic prediction scheme based on the traffic models for these activities is presented. Moreover, they discuss an operation control method to evenly deploy the UAVs. There are two main disadvantages: First, interference is not considered. Second, the operation control method evenly deploys the UAVs, which is not suitable in practice. Reference [27] studies the scenario where a UAV serves UEs which move along a street. It compares with the approaches using stationary BSs and it illustrates that the UAV introduces a considerable gain in channel quality and outperforms the comparing case in throughput. Reference [28] proposes a proactive deployment of cache-enabled UAVs to improve the quality of experience for users. The UAVs cache some popular content based on a prediction model. Such cache would be able to reduce the data packet transmission delay, although achieving the precise prediction model is difficult in real applications. The structure of the system is worth studying. Remote radio heads (RRHs) are grouped into clusters and each RRH can only use the preassigned frequency to avoid interference. RRHs are connected to the cloud of the baseband unit (BBU) via a capacity-constrained DSL link. RRHs transmit data packets to users via cellular links. Drones are connected to BBU through cellular links, and they transmit data packets to users via mmWave. In such a structure, the interference exists in the following two links: RRHs to users and BBU to drones.

The design of UAV communication architecture has also been studied. In [29], the authors discuss two types of communication links: UAV-ground and UAV-UAV. The UAV-ground communication involves control command transmission, UAV status report, UAV-UE data transmission, etc.; and UAV–UAV communication involves sense-and-avoid information-sharing and wireless backhaul support. The UAV-ground communication may have both LoS and NLoS links; while UAV-UAV communication is mainly dominated by the LoS component, so the latest radio access technologies such as mmWave and free-space optical communication can be used [30]. In particular, the UAVs and BSs form a connected graph for the communication purpose, i.e. whenever any served UE sends a request, it can be served without a significant delay due to the connectivity problem. Moreover, Ref. [31] identifies some challenges associated with multitier UAV networks and investigates the feasibility of a two-tier UAV architecture, where some big UAVs are at a higher fixed altitude and small UAVs are at a lower fixed altitude.

The third group of relevant work is about the deployment of small cells. Reference [32] studies how to deploy base stations to satisfy both cell coverage and capacity constraints. Meta-heuristic algorithms are presented to address the optimization problem, where UE density is considered. Reference [33] considers the problem of deploying stationary BSs and small cells. Each small area is associated with a required data rate. The spectral efficiency on bandwidth cannot be lower than a threshold, and some approximation algorithms are proposed. Although these approaches are not for UAV deployment, the basic ideas can be borrowed.

The most relevant sections to this chapter focus on the deployment of UAVs. Reference [5] considers the problem of deploying a single UAV. The objective is to maximize the number of the served UEs, and in the meantime, the users receive acceptable service. Reference [6] solves the problem in [5] by turning the problem into a circle placement problem. Reference [34] focuses on the energy efficiency issue for the downlink with consideration of both the mechanical energy consumption and the communication energy consumption. Reference [35] considers deploying a single UAV with the minimum transmit power to serve a given set of users. Reference [36] investigates the case where a UAV follows a circular trajectory. The center and the radius of the path are adjusted for better performance between ground users and the UAV. Reference [27] studies the scenario where a UAV serves UEs which move along a street. In [37], the optimal position for maximizing the data rate between a stationary BS and a ground user is considered for a single UAV-BS. Additionally, the fading effects are accounted in [38] to find the more accurate position for the UAV-BS to achieve the largest coverage.

There are some other publications focusing on the deployment of multiple UAVs. Reference [7] develops an approach of mapping the UAVs to high traffic demand areas via a neural-based cost function. Reference [13] considers the minimum number of UAVs to cover a given number of users and a heuristic algorithm is proposed. Reference [12] proposes a K-means-based algorithm to find the positions for UAVs that achieve the minimum UAV-user distance. Reference [14] considers the maximum coverage of UEs. The limited working time of UAVs is accounted for therein, and greedy algorithms are proposed. Reference [39] selects the positions of UAVs from a set of candidates to decrease the spatial irregularity of the network, with the presence of BSs. Reference [26] aims at lowering the workload of existing stationary BSs and three typical social activities: stadium, parade, and gathering are considered.

There are some publications in the literature related to the UAV deployment problem [6, 12–14, 16–19, 27, 35, 39–45]. Among them, Refs. [6, 27, 35, 40–42] focus on a single UAV deployment. For example, the paper [40] considers the deployment of a UAV to enhance the communication between a controller and a remote robot. Others consider scenarios with multiple UAVs [12–14, 16–19, 39, 43–46]. In particular, Ref. [39] addresses the problem of selecting UAVs' positions from a set of candidate locations. Reference [16] presents a distributed algorithm to place UAVs guaranteeing a local optimal coverage. The reference [19] considers the issue of the minimum number of UAVs and where to deploy them and designs a particle swarm optimization-based heuristic algorithm such that users receive an acceptable downlink rate. The review [12] studies the user's positioning problem and assigns UAVs to serve users in clusters. In [13], the number of UAVs serving a given set of users is minimized. In [14], the optimal deployment of UAVs to provide wireless service to ground users in urban areas is studied, and the factors including the charging issue and the safety issue are accounted. In [17], the authors consider the problem of using the minimal number of UAVs to completely cover a given area where targets may appear. In [43], the problem of using a given number of UAVs to cover a set of mobile ground users is considered, where the energy constraint is also involved. The paper [18] addresses the problem of maximizing the number of served users by letting UAVs cruise over the area. Reference [46] presents an approach to improve the quality of coverage of targets in an area of interest. The review [44] considers using UAVs to service Internet of things (IoT) devices, which can be active or sleep. In each time slot, the UAVs are repositioned such that the IoT devices can upload data using the minimum power. Furthermore, the trajectories of UAVs are scheduled to minimize energy consumption. One common feature of [6, 12–14, 18, 27, 35, 42–44, 46] is the assumption of knowing the location of each user. Though [45] is based on the spatial probability density function of users, the authors estimate the function based on the prior information about the users' locations. Thus, these publications fall into the category of location-based approaches. However, the users' precise locations may be costly or even impossible to obtain in real-time in practice, thus, this assumption may not always be realistic. In this UAV, a different approach is discussed, in which deploying UAVs is based on UAV-user range.

2.3 UAV-BS Deployment for Maximizing Coverage

In this section, we present the UAV deployment method for the coverage of ground users. We first introduce the system model, based on which we state the problem concerned in Section 2.3.1. In Section 2.3.2, we formally present the developed deployment method.

2.3.1 Problem Statement

We consider a transmission system consisting of n autonomous UAVs labelled $i = 1, 2, \ldots, n$ and k stationary BSs labelled $j = 1, 2, \ldots, k$. The UAVs and BSs should be able to communicate to each other according to a given connected graph C which is called the communication graph. The graph C has $n + k$ vertices $v_1^r, v_2^r, \ldots, v_n^r, v_1^b, \ldots, v_k^b$ that correspond to n UAVs and k BSs. Vertices v_i^r and v_j^r are connected by an edge if the UAVs i and j should be able to communicate. Vertices v_i^r and v_j^b are connected by an edge if the UAV i and the stationary BS j should be able to communicate. Stationary BSs are not required to communicate with each other.

The city environment is described by a so-called "street graph." The street graph S contains M vertices $V_1, V_2, \ldots, V_M$ which are the interceptions of two or more streets. The streets intersect each other only at these vertices. We assume that S is connected. We consider S as a set of points p. We assume that all the stationary BSs $v_1^b, \ldots, v_k^b$ are located at some points $Q_1, \ldots, Q_k$ of S, i.e. $Q_j \in S$ for all j. Furthermore, we deploy all the UAVs at the same altitude in one horizontal plane. Hence, we consider the UAVs' coordinates at that plane as coordinates on the surface. Moreover, we assume that they can be deployed only at some points $P_1, P_2, \ldots, P_n$ of S, i.e. $P_i \in S$ for all i. For any points $p_1, p_2 \in S$, introduce the distance $D(p_1, p_2)$ as the length of the shortest path on S connecting p_1 and p_2; see Figure 2.1. It is obvious that if $E(p_1, p_2)$ is the standard Euclidean distance between p_1 and p_2, then $D(p_1, p_2) \geq E(p_1, p_2)$.[3] Some of our UAVs and stationary

---

3 Notice that the street distance used here will result in inaccuracy when evaluating the communication performance. But as we will show later, such street distance enables us to design an efficient algorithm to deploy UAVs.

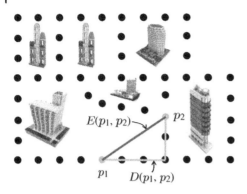

Figure 2.1 An illustrative street graph.

BSs should be close enough to each other to be able to communicate according to C. We will consider deployment of the UAVs satisfying the following constraints:

$$D\left(P_i, P_j\right) \le R_1 \tag{2.1}$$

if the UAVs v_i^r, v_j^r are connected by an edge in C; and

$$D\left(P_i, Q_j\right) \le R_2 \tag{2.2}$$

if the UAV v_i^r and the stationary BS v_j^b are connected by an edge in C. Here $R_1 > 0$ and $R_2 > 0$ are some given constants describing the communication range between our autonomous UAVs and a UAV and a stationary BS. Let $\rho(p) \ge 0$ be a continuous function defined for all $p \in S$. The function $\rho(p)$ is called the density function and describes the density of UE at a particular point p of S at a given time.

Furthermore, let $R > 0$ be a given constant describing the range of high-quality communication from UAVs to UEs. Introduce the set $C(P_1, \dots, P_n) \subset C$ as the set of all points $p \in S$ such that $D(p, P_i) \le R$ for some i. The quality of coverage by the UAVs placed at points $P_1, P_2, \dots, P_n$ is described by the following function:

$$\alpha(P_1, \dots, P_n) := \int_{p \in C(P_1, \dots, P_n)} \rho(p) dp. \tag{2.3}$$

Furthermore, let $c > 0$ and $d > 1$ be given constants, and let $\mathcal{E}$ be the set of pairs (i, j) such that the UAVs i and j are connected by an edge in C. The energy cost of communications between the UAVs i and j located at points P_i and P_j is described by the value $D(P_i, P_j)^d$.[4] Combining this with (2.3), we obtain the following optimization problem:

$$\beta(P_1, \dots, P_n) := \int_{p \in C(P_1, \dots, P_n)} \rho(p) dp - c \sum_{(i,j) \in \mathcal{E}} D(P_i, P_j)^d \to \max, \tag{2.4}$$

where c is the weight between the two terms.

---

4 We assume that UAVs use a fixed power to transmit information to UEs. For the communication between UAVs and BSs, since the data amount from UAVs to BSs can be very small in many cases such as download, where a UE just sends a request to a BS via a UAV, we ignore the data amount from UAVs to BSs. Thus, the communication cost on the side of UAVs is neglected. Further, since BSs have sufficient power supply, the cost on their side is also neglected.

Problem Statement: The optimization problem under consideration is as follows: given the communication graph C, the street graph S, the density function $\rho(p)$, the stationary-based station locations $Q_1, \ldots, Q_k$ and the constants R, R_1, R_2, c, d, find the locations $P_1, \ldots, P_n$ that maximize the function (2.4) subject to the constraints (2.1) and (2.2).

2.3.2 Proposed Solution

The function $\beta(P_1, \ldots, P_n)$ defined for all $P_1, \ldots, P_n \in S$ is a continuous function on a compact set, hence, the maximum in the constrained optimization problem (2.4), (2.1), (2.2) is achieved at some $P_1^0, \ldots, P_n^0 \in S$ [47]. Therefore, there exists a nonempty set of local maximum, and the global maximum is achieved at one of them.

We introduce the following notations: Let $(P_1^0, P_2^0, \ldots, P_n^0)$ be some points on S. Let P_i^0 be an interior point of some edge (V, W) of S connecting the vertices V and W. Then $\mathcal{P}_V(P_i^0)$ and $\mathcal{P}_W(P_i^0)$ denote the sets of points p of the street graph such that $D(p, P_i^0) = R$, $D(p, P_j^0) > R$ for all $j \neq i$, and the shortest path from P_i^0 to p goes in direction to the vertex V and W, respectively. Further, let $\mathcal{E}_V(P_i^0)$ ($\mathcal{E}_W(P_i^0)$) be the set of all j such that the UAVs i and j are connected by an edge in C and the shortest path on S between them goes through the vertex $V(W)$; see Figure 2.2a. Notice that if the shortest path on S between the UAVs i and j is not unique, then $\mathcal{E}_V(P_i^0)$ and $\mathcal{E}_W(P_i^0)$ might have a nonempty intersection. Furthermore, let now P_i^0 be a vertex of S and assume that m edges depart from this vertex, and these edges are labelled $1, \ldots, h$. Then for $h = 1, \ldots, m$, $\mathcal{P}_h(P_i^0)$ denotes the set of points p of S such that $D(p, P_i^0) = R$, $D(p, P_j^0) > R$ for all $j \neq i$, and the shortest path from P_i^0 to p goes through the edge h. Moreover, $\tilde{\mathcal{P}}_h(P_i^0)$ denotes the set of points p of S such that $D(p, P_i^0) = R$, $D(p, P_j^0) > R$ for all $j \neq i$, and the shortest path from P_i^0 to p goes through one of other $m - 1$ edges departing from P_i^0. Finally, let $\mathcal{E}_h(P_i^0)$ be the set of all j such that the UAVs i and j are connected in C and the shortest path on S between the UAVs i and j goes through the edge h, and let $\tilde{\mathcal{E}}_h(P_i^0)$ denote the set of

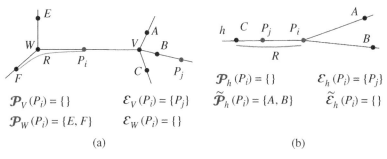

$$\mathcal{P}_V(P_i) = \{\} \qquad \mathcal{E}_V(P_i) = \{P_j\} \qquad \mathcal{P}_h(P_i) = \{\} \qquad \mathcal{E}_h(P_i) = \{P_j\}$$
$$\mathcal{P}_W(P_i) = \{E, F\} \qquad \mathcal{E}_W(P_i) = \{\} \qquad \tilde{\mathcal{P}}_h(P_i) = \{A, B\} \qquad \tilde{\mathcal{E}}_h(P_i) = \{\}$$

          (a)   (b)

Figure 2.2 Illustration of several key notations. (a) P_i is an interior point. (b) P_i is a vertex.

all j such that the UAVs i and j are connected by an edge in C and the shortest path on S between the UAVs i and j goes through one of other $m-1$ edges departing from P_i^0; see Figure 2.2b.

Proposition 2.1 *If* $(P_1^0, P_2^0, \ldots, P_n^0)$ *is a local maximum in the problem (2.4), (2.1), (2.2), then for any* $i = 1, \ldots, n$, *at least one of the following four conditions holds:*

(i) $D(P_i^0, P_j^0) = R_1$ *for some* j *such that the corresponding UAVs* v_i^r, v_j^r *are connected by an edge in* C;

(ii) $D(P_i^0, Q_j) = R_2$ *for some* j *such that the corresponding UAV* v_i^r *and the corresponding stationary BS* v_j^b *are connected by an edge in* C;

(iii) P_i^0 *is an interior point of some edge-connecting vertices* V *and* W *of* S, *and*

$$\sum_{p \in \mathcal{P}_V(P_i^0)} \rho(p) + cd \sum_{j \in \mathcal{E}_V(P_i^0)} D(P_i^0, P_j^0)^{d-1} = \sum_{p \in \mathcal{P}_W(P_i^0)} \rho(p) + cd \sum_{j \in \mathcal{E}_W(P_i^0)} D(P_i^0, P_j^0)^{d-1};$$

$$(2.5)$$

(iv) P_i^0 *is a vertex of* S *with* m *edges departing from it, and*

$$\sum_{p \in \mathcal{P}_h(P_i^0)} \rho(p) + cd \sum_{j \in \mathcal{E}_h(P_i^0)} D(P_i^0, P_j^0)^{d-1} \leq \sum_{p \in \tilde{\mathcal{P}}_h(P_i^0)} \rho(p) + cd \sum_{j \in \tilde{\mathcal{E}}_h(P_i^0)} D(P_i^0, P_j^0)^{d-1}.$$

$$(2.6)$$

Proof. When we are moving P_i^0 along an edge of the street graph, the derivative of the function (2.4) is equal either to the left-hand side of (2.5) minus the right-hand side of (2.5) if P_i^0 is an interior point of an edge, or to the left-hand side of (2.6) minus the right-hand side of (2.6) if P_i^0 is a vertex. Therefore, if all four conditions (i)–(iv) do not hold, we can make a small motion of P_i^0 so that (2.4) increases. Hence, $(P_1^0, P_2^0, \ldots, P_n^0)$ is not a local maximum. □

Let p be some point of S. This point can continuously move along some edge of S in some direction. If p is an interior point of some edge (V, W) of S connecting the vertices V and W, then p can continuously move in two directions, either toward V or toward W. If p be a vertex of the street graph with m edges departing from this vertex, then p can continuously move in m directions along any of these m edges. We propose the following **maximizing algorithm (A1)–(A3)** in TABLE 2.1.

Notice that at some steps P_i cannot move when either (2.5) or (2.6) hold, or beyond P_i at least one of the constraints (2.1), (2.2) ceases to hold.

Proposition 2.2 *The maximizing algorithm* **(A1)–(A3)** *reaches a local maximum of the constrained optimization problem (2.4), (2.1), (2.2) in a finite number of steps.*

Table 2.1 Algorithm A1–A3.

A1: We start with some initial points $(P_1, P_2, \ldots, P_n)$ satisfying the constraints (2.1) and (2.2). Let $N = sn + i$, where $s = 0, 1, 2, \ldots$ and $i = 1, 2, \ldots, n$. At each step N we continuously move the point P_i as follows:

A2: If P_i is an interior point of some edge connecting the vertices V and W of S, and

$$\sum_{p \in P_V(P_i)} \rho(p) + cd \sum_{j \in \mathcal{E}_V(P_i)} D(P_i, P_j)^{d-1} \geq \sum_{p \in P_W(P_i)} \rho(p) + cd \sum_{j \in \mathcal{E}_W(P_i)} D(P_i, P_j)^{d-1} \tag{2.7}$$

then P_i is continuously moving toward V until either it reaches the vertex V, or it reaches a point beyond which at least one of the conditions (2.7), (2.1), (2.2) ceases to hold.

A3: If P_i be a vertex of S with m edges departing from this vertex labeled $1, \ldots, m$, and

$$\sum_{p \in P_h(P_i)} \rho(p) + cd \sum_{j \in \mathcal{E}_h(P_i)} D(P_i, P_j)^{d-1} \geq \sum_{p \in \bar{P}_h(P_i)} \rho(p) + cd \sum_{j \in \bar{\mathcal{E}}_h(P_i)} D(P_i, P_j)^{d-1} \tag{2.8}$$

for some h, then P_i is continuously moving along the edge h until either it reaches another vertex of S, or it reaches a point beyond which at least one of the conditions (2.7), (2.1), (2.2) ceases to hold.

Proof. The algorithm **(A1)–(A3)** generates a sequence $(P_1, P_2, \ldots, P_n)_N$. Since this sequence is in some compact set, it has a subsequence that converges to some $(P_1^0, P_2^0, \ldots, P_n^0)$ [47]. When we are moving P_i^0 along an edge of the street graph, the derivative of the function (2.4) is equal either to the left-hand side of (2.5) minus the right-hand side of (2.5) if P_i^0 is an interior point of an edge, or to the left-hand side of (2.6) minus the right-hand side of (2.6) if P_i^0 is a vertex. Therefore, the function (2.4) is nondecreasing on the sequence $(P_1, P_2, \ldots, P_n)_N$. Hence, $(P_1^0, P_2^0, \ldots, P_n^0)$ is a local maximum. Furthermore, it follows from **(A1)–(A3)** that when $(P_1^0, P_2^0, \ldots, P_n^0)$ in some small neighborhood of **(A1)–(A3)**, it will reach $(P_1^0, P_2^0, \ldots, P_n^0)$ and stays there. □

2.3.3 Evaluation

In this section, we present simulation results for the algorithm **(A1)–(A3)** based on the dataset of Momo. This dataset contains approximately 150 million updates of worldwide users during 38 days, from 21/5/2012 to 27/6/2012 [10]. We extract a subset of it, where the updates come from the users in a 500×500 m^2 residential area in Beijing, China. According to the physical map of this area, we construct a street graph as shown in Figure 2.3. Then, we allocate the extracted updates to the nearest street vertices, based on which we build up the UE density function for each point. All the used parameters can be found in TABLE 2.2.

First, we demonstrate the effectiveness of the algorithm **(A1)–(A3)**. Figure 2.4 shows the relationship between β and N in one simulation with $n = 8$. We can see

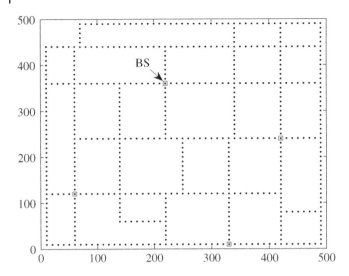

Figure 2.3 The street graph with four stationary BSs.

Table 2.2 Parameter configuration.

Parameter	Value	Description
R	80 m	High-quality communication range from UAVs to UEs
R_1	150 m	Communication range between UAVs and BSs
R_2	150 m	Communication range between UAVs
k	4	Number of BSs
n	5 ~ 15	Number of UAVs
d	2	Parameter for communication cost
c	0.5×10^{-3}	Parameter for communication cost

that β keeps a nondecreasing trend with N and reaches the peak after 11 steps. Second, we consider the selection of c. This parameter can be regarded as the preference of the network supplier. We did simulations for various c with $n = 15$ as shown in Figure 2.5. We can see both the coverage ratio (the ratio of the covered UE number to total number of UEs) and the cost of communication between the UAVs decrease as c increases and when c is larger than 3.5×10^{-3}, they both reach the minimum, because the UAVs cannot be further moved. So in practice, the network supplier can select c considering the trade-off of the coverage ratio and the communication cost. Finally, we demonstrate the results with different

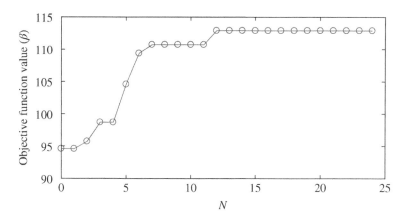

Figure 2.4 Objective function value vs. N.

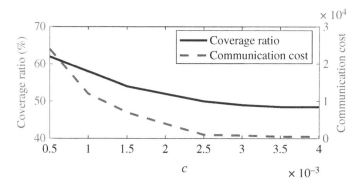

Figure 2.5 The trade-off between the coverage ratio and the communication cost for different values of c.

number of UAVs in Figure 2.6a, b for $c = 0.5 \times 10^{-3}$, i.e. we focus more on the coverage than the communication cost. For each n, we conduct 20 simulations with varying initial deployments. The values of the coverage ratio and the communication cost corresponding to the best β are demonstrated. For comparison, we also show the performance of the corresponding initial deployment of UAVs in each configuration. In Figure 2.6b the communication costs when $n = 5$ and $n = 8$ are zero, because each UAV can directly communicate with BSs. With a larger n, to cover a larger area, some UAVs need to connect BSs via other UAVs, which leads to the communication cost. As seen from Figure 2.6a, b, the proposed algorithm can find better locations for the UAVs compared to their initial positions, which increases the coverage ratio and reduces the communication cost.

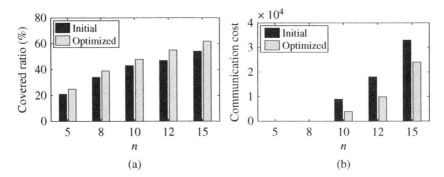

Figure 2.6 Performance for different values of n. (a) Coverage ratio. (b) Communication cost.

2.4 UAV-BS Deployment for Maximizing Coverage and Minimizing Interference

In this section, in addition to the problem of maximizing the coverage of ground users as discussed in Section 2.3, we further consider the influence of interference. In Section 2.4.1, we formulate the problem of interest, and in Section 2.4.2, we present the deployment method.

2.4.1 System Model and Problem Statement

We consider the same system as Section 2.3 consisting of n UAVs labeled $i = 1, 2, \ldots, n$ and k stationary charging poles labeled $j = 1, 2, \ldots, k$. The environment of interest is described by a street graph S, which contains M points $V_1, V_2, \ldots, V_M$. Some of these points are connected by edges. Edges represent streets of the city; see Figure 2.1. We assume that edges intersect each other only at points. We also assume that the street graph S is connected. We consider S as a set of points $p = (x, y)$ of the plane, where (x, y) are the coordinates of p. Let $\rho(p) \geq 0$ be a continuous function defined for all $p \in S$. The function $\rho(p)$ is called the density function and describes the density of UEs at a particular point p of the street graph at a given time. We assume that all the stationary charging poles $v_1^c, \ldots, v_m^c$ are located at some points $Q_1, \ldots, Q_m$ of the street graph, i.e. $Q_j \in S$ for all j. Furthermore, we deploy all the UAVs at the same altitude in one horizontal plane. Hence, we consider the UAVs' coordinates at that plane as coordinates on the surface. Moreover, we assume that they can be deployed only at some points $P_1, P_2, \ldots, P_n$ of the street graph, i.e. $P_i \in S$ for all i. For any points $p_1, p_2 \in S$, $D(p_1, p_2)$ is the length of the shortest path on the street graph connecting p_1 and p_2. It is obvious that if $E(p_1, p_2)$ is the standard Euclidean distance between p_1 and p_2, then $D(p_1, p_2) \geq E(p_1, p_2)$.

The air-to-ground channel model follows [48], which considers LoS and NLoS links. In particular, the LoS probability between a user at $p \in S$ and a UAV at P_i is formulated as follows:

$$P_{LoS}(p, P_i) = \frac{1}{1 + \mu \exp(-v(\tan^{-1}\left(\frac{h}{r(p, P_i)}\right) - \mu))}, \tag{2.9}$$

where $P_{LoS}(p, P_i)$ is the LoS probability, h is the altitude of the UAV, $r(p, P_i)$ is the ground distance between the user and the UAV-BS, and μ and v are environment-dependent constants. As a result of (2.9), the probability of having an NLoS connection can be written as follows:

$$P_{NLoS}(p, P_i^t) = 1 - P_{LoS}(p, P_i^t). \tag{2.10}$$

The path loss (dB) between the user and the UAV can be modeled as follows:

$$PL(p, P_i) = 20 \log\left(\frac{4\pi f_c}{c}\right) + 20 \log(E(p, P_i)) + P_{LoS}(p, P_i)\lambda_{LoS} + P_{NLoS}(p, P_i)\lambda_{NLoS}, \tag{2.11}$$

where f_c is the carrier frequency (Hz), c is the speed of light (m/s), λ_{LoS} and λ_{NLoS} are, respectively, the losses corresponding to the LoS and NLoS connections depending on the environment.

From (2.11), the received signal power at the user $S(p, P_i)$ (watt) can be expressed by:

$$S(p, P_i) = P_t \times 10^{-\frac{PL(p, P_i)}{10}}, \tag{2.12}$$

where P_t (W) is the transmission power of UAV. Further, the interference (W) from other UAVs can be computed in the same way by the following:

$$I(p, P_i) = \sum_{j \in [1, n], j \neq i} P_t \times 10^{-\frac{PL(p, P_j)}{10}}. \tag{2.13}$$

Next, we discuss the models of three main aspects: coverage quality, interference effect, and serving time of UAVs.

First, we consider the modeling of coverage. Let $R_c > 0$ describe the range of high-quality communication from UAVs to UEs. R_c is a surrounding environment-dependent parameter, which can be impacted by such as the altitude of UAVs. We assume that the UAVs work at the same altitude and R_c is identical to all the UAVs. Introduce the set $C(P_1, \dots, P_n)$ as the set of all points $p \in S$ such that $D(p, P_i) \leq R_c$ for some i. If a UAV and the corresponding served users are on the same street, there will be no barriers between the UAV and the users. However, in the urban environment, some served users may be on another street, which is different from the UAV. In this case, the buildings can be regarded as barriers between the UAV and the users. The coverage considered here is based on street distance, and it can handle this case easily.

We define the quality of coverage as the total number of UEs served by the UAVs with high quality. Then, the quality of coverage by the UAVs placed at points $P_1, P_2, \ldots, P_n$ is described by the function (2.3).

Second, we consider the interference model. If the UAVs use different frequency channels to transmit data, as assumed in [12, 13, 26], there will be no interference. However, in most practical cases, the frequency channels are limited and the interference commonly exists. So here we consider the more common situation. A ground UE can receive signals from the serving UAV and nonserving UAVs. The signals from the latter UAVs are the interferences, which challenge the process of signal demodulation. As seen from (2.13), a larger distance from a user and the nonserving UAV means a lower interference. Further, when the distance is large enough, the interference is negligible [18]. Here, we use the following model to describe the interference effect of two UAVs located at points P_i and P_j:

$$f_1(P_i, P_j) = \begin{cases} 0, & \text{if } D(P_i, P_j) \geq R_I; \\ \log(D(P_i, P_j) - R_I + 1), & \text{otherwise.} \end{cases} \quad (2.14)$$

Then, the total interference effect in the system can be represented by the following function:

$$\beta(P_1, \ldots, P_n) := \sum_{i,j \in [1,n], i \neq j} f_1(P_i, P_j). \quad (2.15)$$

Now, we discuss the model related to the serving time of UAVs. We consider the scenario that the UAVs start to hover at certain places to serve UEs with a fully charged battery and when the residual energy reaches a certain level, they fly to the nearest charging pole to recharge the battery. The UAVs rely on the battery for power and the battery is consumed by two basic actions mainly. The first one is hovering at some position to serve UEs, and the second is flying between the hovering position and a charging pole. Although transmitting data to UEs also consumes energy, it can be ignored compared to that on hovering and flying. According to the field experiments using Phantom UAVs [49], the power for flying or hovering is over 140 W; while the typical power for transmitting information through radios is usually around 250 mW [50], which is three orders of magnitude less than the former case. For simplicity, we assume that the energy consumption rate of a UAV is a constant, no matter which mode (hovering or flying) it is in. We also assume that the UAVs fly in the same speed. In this case, the total working time of a UAV is bounded, and the longer time spent on flying means less time on hovering to serve UEs. We introduce R_t to indicate the distance that a UAV can travel between the serving position and the charging pole to guarantee a certain time for serving

UEs. Denote $f_2(P_i, Q_j)$ as the penalty on the distance between a UAV (at P_i) and its charging pole (at Q_j) as follows:

$$f_2(P_i, Q_j) = \begin{cases} 0, & \text{if } D(P_i, Q_j) \le R_t; \\ -(D(P_i, Q_j) - R_t)^c, & \text{otherwise,} \end{cases} \tag{2.16}$$

where $c > 1$. Notice that in graph C, each UAV at P_i is connected to the nearest charging pole, denoted by Q_j.

Then, the overall penalty related to the distance from UAVs and their charging poles is computed as follows:

$$\gamma(P_1, \dots, P_n) := \sum_{i \in [1,n]} f_2(P_i, Q_j). \tag{2.17}$$

Combining (2.3), (2.15,) and (2.17), our objective function is as follows:

$$J(P_1, \dots, P_n) := \int_{p \in C(P_1, \dots, P_n)} \rho(p) dp + a \sum_{i,j \in [1,n], i \ne j} f_1(P_i, P_j) + b \sum_{i \in [1,n]} f_2(P_i, Q_j) \to \max, \tag{2.18}$$

where $a > 0$ and $b > 0$ are two weighting parameters, which make the three terms in (2.18) have the identical unit.

Problem Statement: The optimization problem under consideration is stated as follows: for the given charging graph C, the street graph S, the density function $\rho(p)$, the stationary charging pole locations $Q_1, \dots, Q_m$, and the constants R_c, R_I, R_t, a, b, c, find the locations $P_1, \dots, P_n$ that maximize the function (2.18).

2.4.2 Proposed Solution

In this section, we present a distributed algorithm to address the considered unconstrained optimization problem.

The function $J(P_1, \dots, P_n)$ defined for all $P_1, \dots, P_n \in S$ is a continuous function on a compact set, hence, the maximum in the unconstrained optimization problem (2.18) is achieved at some $P_1^0, \dots, P_n^0 \in S$ [47]. Therefore, there exists a nonempty set of local maximum, and the global maximum is achieved at one of them. This section presents our proposed solution. Before that, we introduce the following assumption and some necessary notations.

We introduce the following notations: Let $(P_1^0, P_2^0, \dots, P_n^0)$ be some points on S. For the point P_i^0, it can be at two types of positions. The first one is at a vertex of S and assume that $l(l > 2)$ edges depart from it, labelled $1, \dots, l$. Notice that if $l = 2, P_i^0$ is at an interior point, which will be discussed below. Then for $h = 1, \dots, l$, $\mathcal{P}_h(P_i^0)$ denotes the set of points P such that $D(P, P_i^0) = R_c, D(P, P_j^0) > R_c$ for all $j \ne i$, and the shortest path from P_i^0 to P goes through the edge h. Further, $\tilde{\mathcal{P}}_h(P_i^0)$ denotes the set of points P such that $D(P, P_i^0) = R_c, D(P, P_j^0) > R_c$ for all $j \ne i$, and the shortest path from P_i^0 to P goes through one of other $l-1$ edges departing from P_i^0. Moreover, $\mathcal{E}_h(P_i^0)$ denotes the set of all j such that the UAVs i and j are within the range of R_I and the shortest path between them is through the edge h, and $\tilde{\mathcal{E}}_h(P_i^0)$ denotes the set of all j such that the UAVs i and j are within the range of R_I and the shortest path between them is via one of other $l-1$ edges. Finally, Q_h denotes the charging pole which is selected by UAV i for charging and the shortest path between them goes through the edge h, and $\tilde{Q}_h$ denotes the charging pole j selected by UAV i and the shortest path between the UAV i and charging pole j goes through one of other $l-1$ edges.

Another case is that P_i^0 is at an interior point of some edge (V, W) of the street graph S connecting the vertices V and W. Then $\mathcal{P}_V(P_i^0)$ and $\mathcal{P}_W(P_i^0)$ denote the sets of points P of the street graph such that $D(P, P_i^0) = R_c$, $D(P, P_j^0) > R_c$ for all $j \neq i$, and the shortest path from P_i^0 to P goes in direction to the vertex V and W, respectively. Further, let $\mathcal{E}_V(P_i^0)$ $(\mathcal{E}_W(P_i^0))$ be the set of all j such that the UAVs i and j are within the range of R_I and the shortest path between the UAVs i and j goes through the vertex $V(W)$. Moreover, we can find the charging pole $Q_v(Q_w)v$ which is the nearest to UAV i via the vertex $V(W)$.

Proposition 2.3 *If $(P_1^0, P_2^0, \dots, P_n^0)$ is a local maximum in the optimization problem (2.18), then for any $i = 1, \dots, n$, at least one of the following two conditions holds:*

(i) P_i^0 is an interior point of some edge-connecting vertices V and W of the street graph S, and

$$\sum_{P \in \mathcal{P}_V(P_i^0)} \rho(P) - \sum_{j \in \mathcal{E}_V(P_i^0)} g_1(P_i^0, P_j^0) + g_2(P_i^0, Q_v) =$$
$$\sum_{P \in \mathcal{P}_W(P_i^0)} \rho(P) - \sum_{j \in \mathcal{E}_W(P_i^0)} g_1(P_i^0, P_j^0) + g_2(P_i^0, Q_w); \tag{2.19}$$

(ii) P_i^0 is a vertex with l edges departing from it, and

$$\sum_{P \in \mathcal{P}_h(P_i^0)} \rho(P) - \sum_{j \in \mathcal{E}_h(P_i^0)} g_1(P_i^0, P_j^0) + g_2(P_i^0, Q_h) \leq$$
$$\sum_{P \in \tilde{\mathcal{P}}_h(P_i^0)} \rho(P) - \sum_{j \in \tilde{\mathcal{E}}_h(P_i^0)} g_1(P_i^0, P_j^0) + g_2(P_i^0, \tilde{Q}_h), \tag{2.20}$$

where $g_1(P_i^0, P_j^0)$ and $g_2(P_i^0)$ are the derivatives of $f_1(P_i^0, P_j^0)$ and $f_2(P_i^0)$, respectively:

$$g_1(P_i^0, P_j^0) = \begin{cases} 0, & \text{if } D(P_i^0, P_j^0) \geq R_I; \\ \frac{a}{\ln(10)} \frac{1}{D(P_i^0, P_j^0) - R_I + 1}, & \text{otherwise;} \end{cases} \tag{2.21}$$

$$g_2(P_i^0, Q_j) = \begin{cases} 0, & \text{if } D(P_i^0, Q_j) \leq R_t; \\ -bc(D(P_i^0, Q_j) - R_t)^{c-1}, & \text{otherwise.} \end{cases} \tag{2.22}$$

Proof. When we are moving P_i^0 along an edge of the street graph, the derivative of the function (2.18) is equal either to the left-hand side of (2.19) minus the right-hand side of (2.19) if P_i^0 is an interior point of an edge, or to the left-hand side of (2.20) minus the right-hand side of (2.20) if P_i^0 is a vertex. Therefore, if all conditions (i)–(ii) do not hold, we can make a small motion of P_i^0 so that (2.18) increases. Hence, $(P_1^0, P_2^0, \ldots, P_n^0)$ is not a local maximum. $\square$

Let P be some point of the street graph. This point can continuously move along some edge of the street graph S in some direction. If P is an interior point of some edge (V, W) of the street graph S connecting the vertices V and W, then P can continuously move in two directions, either toward V or toward W. If P be a vertex of the street graph with l edges departing from this vertex, then P can continuously move in l directions along of any of these l edges. We propose the following **maximizing algorithm B1–B3** in Table 2.3.

Notice that at some steps P_i cannot move when either (2.19) or (2.20) holds. Also, in **B3**, if (2.24) holds for some edge h, we move the UAV along h. Then, the UAV is at an interior point of this edge, which falls into the case of **B2**. So we need to check condition (2.23) to determine whether to move the UAV further or not. In other words, (2.24) turns into (2.23) when the UAV leaves a vertex and moves

Table 2.3 Algorithm B1–B3.

B1: We start with some initial points $(P_1, P_2, \ldots, P_n)$ such that each point is within the range of R_t to the nearest charging pole. At each step $N = sn + i$, where $s = 0, 1, 2, \ldots$ and $i = 1, 2, \ldots, n$, we continuously move the point P_i as follows:
B2: If P_i is an interior point of some edge connecting the vertices V and W of the street graph S, and $$\sum_{P \in \mathcal{P}_V(P_i)} \rho(P) - \sum_{j \in \mathcal{E}_V(P_i)} g_1(P_i, P_j) + g_2(P_i, Q_v) \geq \\ \sum_{P \in \mathcal{P}_W(P_i)} \rho(P) - \sum_{j \in \mathcal{E}_W(P_i)} g_1(P_i, P_j) + g_2(P_i, Q_w); \qquad (2.23)$$ then P_i is continuously moving toward V until either it reaches the vertex V, or it reaches a point beyond which condition (2.23) ceases to hold.
B3: If P_i be a vertex of the street graph with l edges departing from this vertex labeled $1, \ldots, l$, and $$\sum_{P \in \mathcal{P}_h(P_i)} \rho(P) - \sum_{j \in \mathcal{E}_h(P_i)} g_1(P_i, P_j) + g_2(P_i, Q_h) \geq \\ \sum_{P \in \tilde{\mathcal{P}}_h(P_i)} \rho(P) - \sum_{j \in \tilde{\mathcal{E}}_h(P_i)} g_1(P_i, P_j) + g_2(P_i, \tilde{Q}_h), \qquad (2.24)$$ for some h, then P_i is continuously moving along the edge h until either it reaches another vertex of the street graph, or it reaches a point beyond which condition (2.23) ceases to hold.

into the set of interior points of an edge. Further, in **B3** if there are several edges h satisfying (2.24), we choose any of them.

Proposition 2.4 *The algorithm* **(B1)–(B3)** *reaches a local maximum of the optimization problem (2.18) in a finite number of steps.*

Proof. The algorithm **(B1)–(B3)** generates a sequence $(P_1, P_2, \ldots, P_n)_N$. Since this sequence is in some compact set, it has a subsequence that converges to some $(P_1^0, P_2^0, \ldots, P_n^0)$ [47]. When we are moving P_i^0 along an edge of the street graph, the derivative of the function (2.18) is equal either to the left-hand side of (2.19) minus the right-hand side of (2.19) if P_i^0 is an interior point of an edge, or to the left-hand side of (2.20) minus the right-hand side of (2.20) if P_i^0 is a vertex. Therefore, the function (2.18) is nondecreasing on the sequence $(P_1, P_2, \ldots, P_n)_N$. Hence, $(P_1^0, P_2^0, \ldots, P_n^0)$ is a local maximum. Furthermore, it follows from **(B1)–(B3)** that when $(P_1^0, P_2^0, \ldots, P_n^0)$ in some small neighborhood of **(B1)–(B3)**, it will reach $(P_1^0, P_2^0, \ldots, P_n^0)$ and stays there. □

Remark 2.1 In this remark, we briefly discuss the computational complexity and implementation of the algorithm **(B1)–(B3)**. For each move of each UAV, we only need to check one simple inequality, either (2.23) or (2.24). The computational complexity of these inequalities depends on the numbers of elements in the sets $\mathcal{P}_V(P_i)$ and $\mathcal{E}_V(P_i)$. The set $\mathcal{E}_V(P_i)$ consists of UAVs closed to the current one at this moment, so the number of elements in this set cannot be more than $n - 1$. In fact, it is usually much less. The number of elements in the set $\mathcal{P}_V(P_i)$ is a property of the given street graph S. Let s_{max} be the maximum possible number of elements of $\mathcal{P}_V(P_i)$ for the given street graph S. Then the computational complexity of updating by checking the position of each UAV by checking either (2.23) or (2.24) is $O(S_{max}) + O(n - 1)$. Furthermore, since we need to do it for each of n UAVs, the total computational complexity of each step of the algorithm **(B1)–(B3)** is no greater than $O(nS_{max}) + O(n(n - 1)) \leq O(nS_{max}) + O(n^2)$. Such a computational complexity is quite small. Moreover, our investigation of street graphs of major cities shows that in typical situations $S_{max} < 10$. It should also be pointed out that we consider a proactive placement problem, not a reactive placement problem, hence, all the computations required by the algorithm **(B1)–(B3)** are done off-line before the UAVs start their service, so the proposed algorithm does not need real-time computations for its implementation. Therefore, the developed algorithm is of little computational demand and is very easily implementable.

Remark 2.2 In this chapter, we assume that all UAVs are deployed at a constant altitude. In many applications, UAVs can adjust their altitudes to improve coverage. A very interesting direction for future research will be extending the approach of this chapter to the 3D UAV deployment problem. In this case, the UAVs altitude z is taking values in some interval $z \in I_z = [Z_{min}, Z_{max}]$, and UAV can be deployed in the set $S \times I_z$. The density function $\rho(p, z)$ now depends on the altitude z as well and describes the density of UEs accessible from a given altitude. Furthermore, the functions $f_1(P_i, P_j, z)$ and $f_2(P_i, z)$ from (2.18) also depend on z. The inequalities (2.23) and (2.24) should now also include the partial derivatives of $\rho(p, z), f_1(P_i, P_j, z)$ and $f_2(P_i, z)$ with respect to z. Instead of moving P_i along an edge h of the street graph as in the algorithm **(B1)–(B3)**, we now need to move P_i in the set $h \times I_z$ building a trajectory such that the modified inequality (2.23) holds. Since the amount of computations becomes much bigger, an efficient computational algorithm should be developed.

2.4.3 Simulation Results

This section presents simulation results for the algorithm **(B1)–(B3)** based on the dataset Momo.

2.4.3.1 Dataset and Simulation Set-Up

For our research, we extract a subset of the Momo dataset, where the records come from the UEs in a $1000 \times 1000 \, \text{m}^2$ residential area in Beijing, China. We construct a street graph with the size of $M = 500$, as shown in Figure 2.7. Then, we allocate the extracted updates to the nearest street points, based on which we build up the UE density function for each point. The UE density is further normalized into the

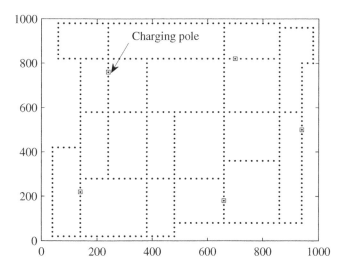

Figure 2.7 The street graph with five stationary charging poles.

range from 0 to 1. Notice that although the UE density function built upon Momo dataset cannot describe the real UE distribution since only a part of UEs use this App, while many others do not use it, it is still better than the random distribution used in many existing publications. Further, we assume there are five charging poles located on the street graph as shown in Figure 2.7.

All the parameters used in our simulations can be found in Table 2.4. Notice that in practice, most of these parameters, such as R_c, should be obtained based on field experiments. The performance metric considered here is the objective function value J and when we go deep into the influence of R_c, R_I, and R_t, we further observe the coverage ratio (the number of UEs served with high quality to the total number of UEs in the considered area) and the interference effect.

2.4.3.2 Comparing Approaches

As discussed in Section 2.2, there are some existing publications on the topic of UAV deployment. However, some consider the deployment of a single UAV [5, 6, 27, 34–36]; some are based on the exact users' locations [8, 12, 13]; and others consider different aspects, for example, Ref. [7] targets at assigning UAVs to high dense areas and [39] aims at selecting UAVs' locations for maximizing the sum of the distance between the UAVs and BSs. Although all these related works are on the same topic as this chapter, they consider various objectives and are based on different conditions. With this regard, we compare our distributed algorithm with a greedy algorithm [14]. The basic idea of such a greedy algorithm is that in each step, it looks for the position for one UAV such that the objective function (2.18) is maximized. Clearly, this algorithm is centralized.

2.4.3.3 Simulation Results

The performance of our algorithm depends highly on the initial condition and various initial conditions lead to different results. Thus, it is easy to observe that the more simulations we conduct, the better positions for UAVs can be found.

First, we demonstrate the effectiveness of the proposed algorithm using one simulation result ($n = 15$). As seen in Figure 2.8, every movement of a UAV leads

Table 2.4 Parameter configuration.

Parameter	Value	Parameter	Value
R_c	200 m	R_I	1000 m
R_t	500 m	n	$3 \sim 15$
k	5	a	0.5
b	1	c	2

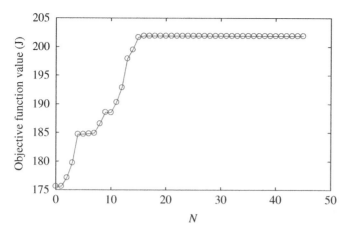

Figure 2.8 Objective function value vs. *N* (*n* = 15).

to a rise of the objective function or remaining steady, which further illustrate the convergence of the algorithm **(B1)–(B3)** shown in Proposition 2.4. Compared to the initial positions, the final positions increase the objective function value by 15%. The objective function reaches the local maximum after 17 steps.

Second, we consider the trade-off of coverage ratio (the ratio of the covered UE number to the total number of UEs) and interference effect, because they are competing with each other. Here, we study the performance for different *a* values. We conduct simulations for various *a* with *n* = 15, and these simulations are with the same initial positions. The results are summarized in Figure 2.9. We can see

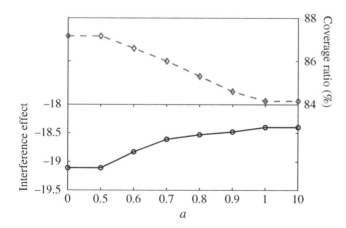

Figure 2.9 The trade-off between the coverage ratio and the interference effect for different values of *a*.

that when a is smaller than 0.5, more attention is paid to the coverage ratio so that the coverage ratio achieves the maximum; while the interference effect is the strongest. When a is larger than 1.0, the weak interference is more preferred, so that the interference effect is the weakest; while the coverage ratio is the lowest. When a is between 0.5 and 1.0, the coverage ratio decreases as a increases while the interference effect becomes weaker as a increases. So in practice, the network supplier can select a considering the trade-off of the coverage ratio and the interference effect. In the below simulations, we set a as 0.5.

Now, we display the simulation results for different numbers of UAVs in Figure 2.10. The first one is the best result among 50 sets of initial positions; while the second is among 100 sets. We can see that with a larger number of trials, our approach can find better positions for the UAVs and outperforms the greedy algorithm. The main reason is that our algorithm manipulates several UAVs together to better serve the area, while the greedy algorithm seeks the current largest benefit in each step without considering the overall performance. Moreover, it is interesting to study how far locally optimal solutions obtained by the proposed method are from globally optimal solutions. Therefore, we have conducted finding the global optimum by the brute-force search (exhaustive search). More precisely, we checked possible positions of UAVs satisfying the constraints on the street graph with some very small steps. The computing demand of such an exhaustive search algorithm increases exponentially with the number of UAVs; therefore, finding the global optimum by it is possible only for small numbers of UAVs. We present the global optimums for three and five UAVs in Figure 2.10. It can be seen that when $n = 3$, the results of the proposed algorithm for both 50 and 100 trials are the same as the global optimum; while for $n = 5$, the results are very close to the global optimum. Of course, for larger n,

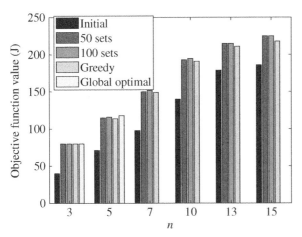

Figure 2.10 Comparison of the proposed algorithm with the greedy algorithm. Notice that the "Initial" is the result corresponding to the 50 sets and that of the 100 sets is not shown.

the brute-force search requires a huge amount of computations; hence, it cannot be implemented.

Moreover, we investigate the influence of several key parameters. To get a deep understanding of their impact, when we observe the influence of one parameter, we fix the others. For this purpose, we summarize four groups of results with the same initial positions under various configurations of three important parameters: R_c, R_I, and R_t. The standard configuration of them is the same as above, i.e. $R_c = 200$, $R_I = 1000$, and $R_t = 500$. In each of the other three configurations, we only change one parameter against the standard one. So in Figure 2.11a, b, there are three groups of comparisons between the standard configuration with the other three configurations. Here, we mainly focus on the coverage ratio and the interference effect. First, we consider the impact of R_c. As mentioned earlier, R_c depends on the environment. Also, different countries have certain regulations on the maximum height of UAVs.[5] Here, we compare the performance under different values of R_c. Comparing the second configuration with the standard one, we can clearly see from Figure 2.11a that larger R_c leads to a higher coverage ratio since the coverage area of a single UAV is increased. However, larger R_c worsens the interference effect, see Figure 2.11b. The reason behind this is that the contribution of the quality of coverage to the objective function is increased, while the interference effect is sacrificed. As for R_I, i.e. comparing the third configuration with the standard one, we can see that smaller R_I results in larger coverage ratio (see Figure 2.11a) and lower interference effect (see Figure 2.11b). The reason is that when R_I becomes smaller, according to (2.14), the interference effect is reduced under the same distance, then UAVs are more likely to find better positions to improve the coverage quality. R_t is a parameter relating to the size of the battery

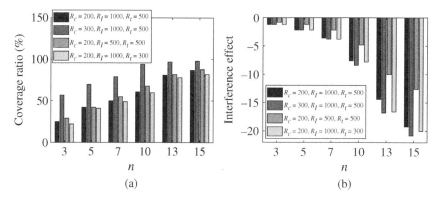

Figure 2.11 The influence of R_c, R_I, and R_t on (a) coverage ratio, (b) interference effect.

---

5 https://bit.ly/2lAT0ep; https://bit.ly/2KlLyhF; https://bit.ly/21dvK3i.

UAVs can carry. But according to the regulations provided in Footnote 3, different countries also have various constraints on the weight of UAVs. Generally, the larger the allowed weight, the longer the UAVs can travel and then the longer R_t can be. Here, we compare the performance under different values of R_t. As shown in Figure 2.11a, smaller R_t reduces the coverage ratio and its impact on the interference effect is a negative trend, especially under large n, see Figure 2.11b. This is because that when R_t becomes smaller, under the same distance from hovering position to charging pole, the penalty becomes larger. The range of candidate positions of UAVs is narrowed to avoid being penalized, which contributes to the drop in coverage quality. Also, it raises the interference effect, because the relative distance between two UAVs, which select the same charging pole, becomes smaller.

To sum up, via simulations, in this section, we illustrated the convergence of the proposed algorithm, studied the trade-off between maximizing coverage and minimizing interference, compared our algorithm with a greedy one and investigated the impact of several key parameters on the performance.

2.5 Voronoi Partitioning-Based UAV-BS Deployment

In Sections 2.3 and 2.4, we discussed how to deploy UAV-BSs in a street graph. In this section and the following one, we remove this constraint and consider the more general deployment problem. In this section, we present a Voronoi partitioning-based deployment approach, and in the following section, we present a range-based deployment method.

2.5.1 Problem Statement and Main Results

During some special occasional event, lots of users enter some areas. Due to the large demand, existing SBSs cannot provide service to users in the area. Thus, we deploy UAV-BSs to provide service to users during such events, see Figure 2.12a.

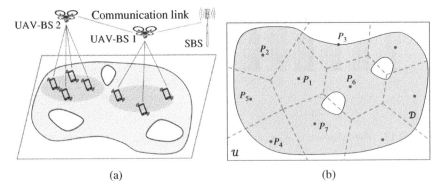

(a)   (b)

Figure 2.12 (a) Illustration of the considered scenario. (b) Construction of Voronoi cells.

We consider a transmission system consisting of n UAV-BSs labeled $i = 1, 2, \ldots, n$ and k SBSs labelled $j = 1, 2, \ldots, k$. Some UAV-BSs and SBSs should be able to communicate to each other according to a given connected graph C which is called the communication graph. The vertices $v_1^r, v_2^r, \ldots, v_n^r$ represent the UAVs, and $v_1^b, \ldots, v_k^b$ correspond to k SBSs. Vertices v_i^r and v_j^r are connected by an edge if the UAV-BSs i and j should be able to communicate. Vertices v_i^r and v_j^b are connected by an edge if the UAV-BS i and the SBS j should be able to communicate. SBSs are not required to communicate to each other. For instance, in Figure 2.12a, UAV-BS 2 is connected to UAV-BS 1, and UAV-BS 1 is also connected to the working SBS.

We deploy all the UAV-BSs at the same altitude[6] in one horizontal plane [12, 13]. Hence, we consider the UAV-BSs' coordinates at that plane as coordinates on the surface. Moreover, let D be some given bounded, closed set with smooth boundary which is called the UAV-BS deployment region. We assume that they can be deployed only at some points $P_1, P_2, \ldots, P_n$ of this region, i.e. $P_i \in D$ for all i. Notice that D may have "holes" that may correspond, for example to tall buildings or some other areas over which we are not allowed to deploy UAV-BSs.

Considering the limitation of communication range, some of our UAV-BSs and SBSs should be close enough to each other to be able to communicate according to the communication graph $\mathcal{G}$. We will consider the deployment of UAV-BSs satisfying the following constraints (2.1) and (2.2).

Let (x, y) be the standard Cartesian planar coordinates. The considered area is described by a bounded, closed region $\mathcal{U}$[7] and a continuous function $\rho(x, y) \geq 0$ defined for all $(x, y) \in \mathcal{U}$, that are called the user region and the user probability density function, correspondingly. The function $\rho(x, y)$ describes the probability density of users at a particular point (x, y) of the user region at a given time and satisfies the standard probability constraint $\int \int_{\mathcal{U}} \rho(x, y) dx \, dy = 1$. Further, we introduce the function $r(x, y) \geq 0$ to describe the data rate requirement of the users at point (x, y); and we introduce another function $f(x, y) := r(x, y)\rho(x, y)$ as the weighted user probability density.

Now, let $P_1, P_2, \ldots, P_n$ be some given locations of UAV-BSs, and introduce the function $d_{min}(x, y)$ defined for all $(x, y) \in \mathcal{U}$ as follows:

$$d_{min}(x, y) := \min_{i=1,\ldots,n} |(x, y), P_i|, \tag{2.25}$$

i.e. $d_{min}(x, y)$ is the distance from the point (x, y) to the nearest UAV-BS.

The performance of the UAV-BSs placed at points $P_1, P_2, \ldots, P_n$ is described by the following function:

$$q(P_1, \ldots, P_n) := \int \int_{\mathcal{U}} d_{min}^2(x, y) f(x, y) dx \, dy. \tag{2.26}$$

---

6 This assumption is valid when we have lower and upper bounds for the altitudes of UAV-BSs, so that the space is narrow. The case with varying altitudes is left for future study.

7 $\mathcal{U}$ is exclusive of some areas, where SBSs can provide service to users.

In other words, the function $q(P_1, \dots, P_n)$ is the mathematical expectation $E[d^2_{min}]$ of the squared distance between a user and the nearest UAV-BS. Now, we consider the following optimization problem:

$$\min_{P_1,\dots,P_n} q(P_1, \dots, P_n), \tag{2.27}$$

where the minimum is taken over all $P_1, \dots, P_n \in \mathcal{D}$ satisfying the constraints (2.1) and (2.2).

We define the quality of coverage achieved by the UAV-BSs as $1/q(P_1, \dots, P_n)$. It is clear that minimizing (2.27) is equivalent to maximizing the quality of coverage.

Problem Statement: The optimization problem (2.27) under consideration is as follows: for the given communication graph $\mathcal{G}$, the UAV-BS deployment region $\mathcal{D}$, the user region $\mathcal{U}$, the probability density function $f(x, y)$, the SBSs' locations $Q_1, \dots, Q_k$ and the constants r_1, r_2, find the locations $P_1, \dots, P_n \in \mathcal{D}$ that minimize the function (2.26) subject to the constraints (2.1), (2.2).

The function $q(P_1, \dots, P_n)$ defined for all $P_1, \dots, P_n \in \mathcal{D}$ is a continuous function on a compact set, hence, the minimum in the constrained optimization problem (2.27), (2.1), (2.2) is achieved at some $P^0_1, \dots, P^0_n \in \mathcal{D}$ [47]. Therefore, there exists a nonempty set of local minima, and the global minimum is achieved at one of them.

Notice that the problem (2.27), (2.1), (2.2) is different to the problem of deployment of SBSs; see e.g. [33] and references therein. First, the latter problem generally considers the location constraints. In practice, SBSs can be deployed to a small number of candidate locations, thus integer programming is usually used to model the problem. While in our model, the candidate positions of UAV-BSs are continuous. Further, the backhaul connection is not a serious issue in deploying SBSs, while it is critical in UAV-BSs, because SBSs can connect the core network through the underground cables, while wireless connectivity is necessary for UAV-BSs.

We will need the following notations. For any $i = 1, 2, \dots, n$, the Voronoi cell $C_i(P_1, \dots, P_n)$ is defined as the set of points $(x, y) \in \mathcal{U}$ such that $|(x, y), P_i| \leq |(x, y), P_j|$ for all $j = 1, 2, \dots, n$, see Figure 2.12b. Moreover, the point $(X_i(P_1, \dots, P_n), Y_i(P_1, \dots, P_n))$, where

$$X_i(P_1, \dots, P_n) := \frac{\int\int_{C_i(P_1,\dots,P_n)} xf(x,y)dx\,dy}{\int\int_{C_i(P_1,\dots,P_n)} f(x,y)dx\,dy},$$

$$Y_i(P_1, \dots, P_n) := \frac{\int\int_{C_i(P_1,\dots,P_n)} yf(x,y)dx\,dy}{\int\int_{C_i(P_1,\dots,P_n)} f(x,y)dx\,dy} \tag{2.28}$$

is called the center of the Voronoi cell $C_i(P_1, \dots, P_n)$. Notice that the center point $(X_i(P_1, \dots, P_n), Y_i(P_1, \dots, P_n))$ may be outside of $C_i(P_1, \dots, P_n)$ if $C_i(P_1, \dots, P_n)$ is not convex.

Definition 2.1   Let $P_1, \ldots, P_n$ be some locations of UAV-BSs. A location P_i is said to be a boundary point if either $|P_i, P_j| = r_1$ for some j such that the corresponding UAV-BSs v_i^r, v_j^r are connected by an edge in $\mathcal{G}$, or $|P_i, Q_j| = r_2$ for some j such that the corresponding UAV-BS v_i^r and the corresponding SBS v_j^b are connected by an edge in $\mathcal{G}$, or P_i belongs to the boundary of the UAV-BS deployment region $\mathcal{D}$.

Proposition 2.5   *If $(P_1^0, P_2^0, \ldots, P_n^0)$ is a local minimum in the constrained optimization problem (2.27), (2.1), (2.2), then for any $i = 1, \ldots, n$, P_i^0 is either the center of the Voronoi cell $C_i(P_1^0, \ldots, P_n^0)$ or a boundary point.*

Proof.   It obviously follows from the definition of the Voronoi cell $C_i(P_1, \ldots, P_n)$, (2.25) and (2.26) that

$$q(P_1, \ldots, P_n) := \sum_{i=1}^{n} \int \int_{C_i(P_1^0, \ldots, P_n^0)} |(x, y) - P_i^0|^2 f(x, y) dx\, dy. \tag{2.29}$$

Furthermore, let (x_i^0, y_i^0) be the coordinates of P_i^0, then

$$\int \int_{C_i(P_1^0, \ldots, P_n^0)} |(x, y) - P_i^0|^2 f(x, y) dx\, dy =$$
$$\int \int_{C_i(P_1^0, \ldots, P_n^0)} ((x - x_i^0)^2 + (y - y_i^0)^2) f(x, y) dx\, dy =$$
$$a((X_i(P_1^0, \ldots, P_n^0) - x_i^0)^2 + (Y_i(P_1^0, \ldots, P_n^0) - y_i^0)^2) + c =$$
$$a|O_i - P_i^0|^2 + c, \tag{2.30}$$

where

$$a := \int \int_{C_i(P_1^0, \ldots, P_n^0)} f(x, y) dx\, dy,$$
$$c := \int \int_{C_i(P_1^0, \ldots, P_n^0)} (x^2 + y^2) f(x, y) dx\, dy - a(X_i(P_1^0, \ldots, P_n^0)^2 + Y_i(P_1^0, \ldots, P_n^0)^2),$$

$$\tag{2.31}$$

and O_i is the center of the Voronoi cell $C_i(P_1, \ldots, P_n)$, $O_i(X_i(P_1^0, \ldots, P_n^0)$, $Y_i(P_1^0, \ldots, P_n^0))$. We now prove that if $(P_1^0, P_2^0, \ldots, P_n^0)$ is a point of local minimum and P_i^0 is not the Voronoi cell's center O_i, then P_i^0 is a boundary point. Indeed, consider the straight line segment connecting P_i^0 and O_i. It obviously follows from (2.29), (2.30) that when we move the point P_i along this segment from P_i^0 to O_i, the cost function (2.26) is strictly decreasing. If P_i^0 is not a boundary point, then for some small movement of the point P_i^0 toward O_i, the constraints (2.1) and (2.2) will still hold, and P_i^0 will still be in $\mathcal{D}$. This means that, we can make a small motion of P_i^0 so that the function (2.27) strictly decreases, and all

Table 2.5 Algorithm C1–C3.

C1: We start with some initial points $P_1^{(0)}, \ldots, P_n^{(0)} \in D$ satisfying the constraints (2.1) and (2.2).

C2: At each step $t = 1, 2, \ldots$, for any $i = 1, \ldots, n$, we calculate a new point $P_i^{(t)}$ as follows: let $O_i := (X_i(P_1^{(t-1)}, \ldots, P_n^{(t-1)}), Y_i(P_1^{(t-1)}, \ldots, P_n^{(t-1)}))$ the center of the Voronoi cell $C_i(P_1^{(t-1)}, \ldots, P_n^{(t-1)})$.

C3: If $O_i \in D$ and the constraints (2.1), (2.2) with P_i replaced by O_i hold, then we take $P_i^{(t)} := O_i$. Otherwise, $P_i^{(t)}$ is the boundary point closest to O_i.

---

the constraints still hold. Hence, $(P_1^0, P_2^0, \ldots, P_n^0)$ is not a local minimum. This completes the proof of Proposition 2.5. □

We propose the recursive algorithm minimizing the mathematical expectation (2.26) shown in Table 2.5. Notice that we evaluate the quality of coverage q in each step t and when the difference of q in two continues steps is lower than a given threshold ε, we say the positions of UAV-BSs converge to the locally optimal positions. In **C3**, if there are more than one boundary points closest to O_i, we take among them the point closest to $P_i^{(t-1)}$. Besides, when we determine a new position for a UAV-BS, we only consider the nearby UAV-BSs to compute the Voronoi cell, rather than all the UAV-BSs. In the example of Figure 2.12b, only six UAV-BSs influence the Voronoi cell of P_1 and only three UAV-BSs influence the Voronoi cell of P_5, although there are totally 10 UAV-BSs. In other words, this algorithm is scalable to a large number of UAV-BSs, and thus, it is superior to some existing centralized methods. We end this section by showing the convergence of our algorithm.

Proposition 2.6 *The algorithm **C1–C3** converges to a local minimum of the constrained optimization problem (2.27), (2.1), (2.2).*

Proof. It obviously follows from (2.38) and (2.39) that if $|P_i - O_i| \geq |\hat{P}_i - O_i|$, where O_i is the center of the Voronoi cell $C_i(P_1, \ldots, P_n)$, then $q(P_1, \ldots, P_i, \ldots, P_n) \geq q(P_1, \ldots, \hat{P}_i, \ldots, P_n)$. Therefore, $q(P_1^{t-1}, \ldots, P_n^{t-1}) \geq q(P_1^t, \ldots, P_n^t)$ for all $t = 1, 2, \ldots$. Moreover, since the sequence $(P_1^t, \ldots, P_n^t)$ is in some compact set D, it has a subsequence that converges to some $(P_1^0, P_2^0, \ldots, P_n^0)$ [47]. Now, it follows from Table 2.5 and the fact that $q(P_1^t, \ldots, P_n^t)$ is decreasing, that $(P_1^0, P_2^0, \ldots, P_n^0)$ is a local minimum and the sequence $(P_1^t, \ldots, P_n^t)$ converges to it. This completes the proof of Proposition 2.6. □

At each step, the algorithm involves checking conditions (2.1), (2.2), building Voronoi cells of UAV-BSs current locations, and calculating the center of mass of the built Voronoi cells according to formula (2.28). The computational complexities of these operations are $O(n^2)$, $O(nk)$, $O(n^2)$ and $O(n)$, correspondingly. Hence, the computational complexity of each step of the proposed algorithm is $O(n^2 + nk)$.

2.5.2 Simulation Results

We present simulation results based on the real dataset Momo. Momo is a social discovery App, and when a Momo user has an update, the information of his ID, timestamps, latitude, and longitude is sent to the server. This dataset contains approximately 150 million world-wide users' updates in 38 days. We only focus on the updates by the users in a $1200 \times 800\,\text{m}^2$ area. For the timestamps, we consider two one hour periods. We name them the peak hour and the less peak hour, respectively. The distributions of users during these two periods are similar, but the average number of users in the peak hour (1399) is larger than that in the less peak hour (948). Further, we construct a plane with the grid size of 10 m and the user probability density is the ratio of the average number of users that fall into each cell to the total number of users; see Figure 2.13a. We use colors to indicate the user probability, and the lighter the color, the larger the probability is. Also, we construct a UAV-BS deployment region $\mathcal{D}$. It has three holes, where we are not allowed to deploy UAV-BSs. There are four SBSs located at the corners (marked by black $\times$), and their communication ranges are shown by the black lines in Figure 2.13a. $r_1 = 300\,\text{m}$, and $r_2 = 250\,\text{m}$. $r(x, y) = 1, \forall (x, y) \in \mathcal{U}$. Some areas on the four corners are covered by SBSs (Figure 2.13a).

We conduct simulations by MATLAB for the peak and less peak hour cases, respectively. We demonstrate the movements of 15 UAV-BSs for the peak hour case in Figure 2.13a. From the initial positions (marked by circles), they move along the solid lines. For the final positions, we further plot the Voronoi cells,

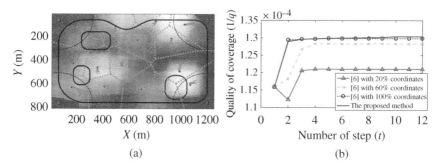

Figure 2.13 (a) The trajectories of 15 UAV-BSs and the final Voronoi cells for the peak hour. (b) The quality of coverage vs. the number of steps for the peak hour by the proposed algorithm and the one in [12]. Source: Savkin and Huang [22].

marked by the dashed lines. Correspondingly, we show the performance in the quality of coverage against the number of steps (t) in Figure 2.13b. We can see that the quality of coverage increases by about 13% in 12 steps, with $\varepsilon = 10$. Moreover, we compare our algorithm with the method of [12]. The algorithm of [12] is based on knowing users' locations which are very difficult in practice. To make it work in our scenario, the grids are treated as users, and the densities are the numbers of users. Considering the difficulty of obtaining the coordinates of users, especially, when in events with a huge number of users (e.g. over one million), we consider cases where only a part of coordinates of users are known to the algorithm [12]. The percentages are taken as 20%, 60%, and 100%. Starting from the same initial positions as shown in Figure 2.13a, the results in Figure 2.13b show that when only 20% or coordinates are known, the performance of [12] drops first and then increases with t. With the increase of the percentage of users' coordinates, the performance of [12] improves, and it achieves competitive results with our algorithm only when all the users' coordinates are known, which may be impossible in many real applications. Therefore, the proposed algorithm significantly outperforms the method of [12] if we do not assume that all users' coordinates are precisely known.

Furthermore, we show the simulation results for various numbers of UAV-BSs. Figure 2.14a shows how the quality of coverage increases with n. As the peak hour and the less peak hour case have similar user distributions, the corresponding normalized user probability density functions are also similar. Moreover, since each UAV-BS has limited bandwidth, we assume that there is an upper bound for the number of users a UAV-BS can serve. We set this bound be equal to 130. We plot the coverage ratio[8] for the two cases in Figure 2.14b. We can see that to fully cover all the users, we need at least 10 UAV-BSs during the less peak hour, while we need at least 14 UAV-BSs during the peak hour.

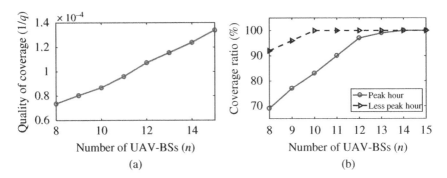

Figure 2.14 (a) The quality of coverage VS the number of UAV-BSs. (b) The coverage ratio VS the number of UAV-BSs.

---

8 The ratio of the effectively served users to the total number of users in $\mathcal{U}$.

2.6 Range-Based UAV-BS Deployment

Following Section 2.5, we consider the scenario where UEs are on a 2D region, and we present a UAV-BS deployment method based on the range measurement only. Different from the Voronoi partitioning-based approach discussed in Section 2.5, the method presented in this chapter does not require the location information of the ground users. Instead, only the range information is needed.

2.6.1 Problem Statement and Main Results

Consider a network of autonomous UAVs consisting of n UAVs labeled $i = 1, 2, \ldots, n$. Following [13], the UAVs are backhaul-connected via satellite links. Each UAV i is deployed at some given altitude $a_i > 0$ in the corresponding horizontal plane. The altitudes a_i are selected to avoid physical collisions between UAVs. Furthermore, let $D_1, D_2, \ldots, D_n$ be the UAVs' coordinates in that plane as coordinates on the surface. For any UAV i, $D_i = (D_i^x, D_i^y)$.

There are N users labeled $1, 2, \ldots, N$ with the coordinates $U_1, U_2, \ldots, U_N$, where $N > n$, $U_j \in \mathbf{R}^2$. First, it is assumed that the users are stationary, but their coordinates are unknown. Let d_{ij} denote the standard Euclidean distance between the UAV i and the user j. Then, d_{ij} can be computed by the following:

$$d_{ij} = \sqrt{|D_i, U_j|^2 + a_i^2},\tag{2.32}$$

where $|\cdot, \cdot|$ denotes the standard Euclidean distance between two points. Furthermore, for all $j = 1, 2, \ldots, N$, introduce the distance $\tilde{d}_j$ as follows:

$$\tilde{d}_j := \min_{i=1,\ldots,n} d_{ij}.\tag{2.33}$$

In other words, $\tilde{d}_j$ is the distance from the user j to its closest autonomous UAV. The following notations are needed. The set of N users is divided into n nonintersecting subsets $\mathcal{U}_1, \mathcal{U}_2, \ldots, \mathcal{U}_n$, where for any $i = 1, 2, \ldots, n$, $\mathcal{U}_i$ is the set of users j such that $\tilde{d}_j = d_{ij}$, i.e. the set of users for which the UAV i is the closest one at this given time. In cases, when the shortest distance $\tilde{d}_j$ is achieved with more than one UAV, the user j belongs to any one of all such UAVs. Notice that the sets $\mathcal{U}_i$ might change with time as the UAVs move and the users might have different closest UAVs at different times.

The locations (coordinates) of the users are not known. However, the UAVs know distances from themselves to nearby users. More precisely, each UAV i at any time knows the distances d_{ij} for all $j \in \mathcal{U}_i$. In other words, each UAV at any time knows the distances to the users for which this UAV is the closest one. In practice, it is achieved by estimation of the distance d_{ij} from the user signal strength and information exchange with nearby UAVs which allows to find out whether for a particular user this UAV is the closest one or not. There are some schemes to estimate the distances in the literature, such as the robust extended Kalman filter [51]. Consider the case where UAV i collects signals from user j. Let p_{ij} denote the received signal strength, which can be expressed as a function of distance d_{ij}:

$$p_{ij} = p_{oi} - 10\epsilon \log(d_{ij}) + v_i, \tag{2.34}$$

where p_{oi} is a given constant determined by the transmitted power, ϵ is a slope index depending on the environment, and v_i is the logarithm of the shadowing component, which is regarded as an uncertainty in the measurement. If we replace d_{ij} by Eq. (2.32), we can further obtain the relationship between p_{ij} and the positions of the user and the UAV. When we measure the received signal strength p_{ij}, the corresponding distance d_{ij} can be estimated following the methods discussed in [52, 53]. Since this is not a key point in this chapter, the corresponding technical details are not included. Readers can refer to [52, 53] and the references therein for more details.

Following [12, 13], it is assumed that the UAV-user communication channels are dominated by LoS,[9] and the main factor affecting the quality of coverage is pathloss. Thus, maximizing the quality of coverage that a UAV can provide to a user is equivalent to minimizing the UAV-user distance. Notice that as improving the quality of coverage and simultaneously reducing the interference is a very challenging issue in deploying UAVs, in this chapter, we do not focus on reducing interference but only consider how to improve the quality of coverage. For the users at $U_1, U_2, \ldots, U_N$, the quality of coverage by the UAVs placed at points $D_1, D_2, \ldots, D_n$ is described by the following function:

$$W(D_1, \ldots, D_n) := \frac{1}{N} \sum_{j=1,\ldots,N} \tilde{d}_j^2. \tag{2.35}$$

Now, consider the following optimization problem:

$$W(D_1, \ldots, D_n) \to \min, \tag{2.36}$$

where the minimum is taken for fixed users' locations $U_1, U_2, \ldots, U_N$ over all possible UAV locations $D_1, \ldots, D_n$. In other words, the objective is to find UAVs' locations that minimize the average squared distance from users to their closest UAVs.

Problem Statement: The optimization problem under consideration is as follows: for the given users' locations $U_1, U_2, \ldots, U_N$ (not known to the UAVs a priori), and the UAV altitudes $a_1, a_2, \ldots, a_n$, move the UAVs to locations $D_1, D_2, \ldots, D_n$ that minimize the function (2.36).

---

9 Besides serving the affected users who are moved to some open areas due to disasters, another scenario supporting the LoS assumption is livestock monitoring such as cattle tracking [54, 55]. When UAVs are employed to collect data of the animals, LoS is usually available.

Remark 2.3 A distinctive feature of the proposed approach is that it is based on distances from UAVs to users and does not require information about users' locations. Users' locations can be estimated from the received signal strength using the robust extended Kalman filter [52, 53]. However, it may not be necessary to do this. Under the considered analytical model with the basic assumption of LoS links, no matter where a user is, as long as the distance between the user and the serving UAV stays the same, the user can receive the same quality of service. In other words, the quality of service depends on the relative distance between the UAV and the user, rather than the relative position. Furthermore, estimating users' locations would increase the total communication and/or computation loads of the system, which is not necessary.

Let $\mathcal{V}$ be the convex hull of the users' locations $U_1, \dots, U_N$. It is obvious that in search of $D_1, \dots, D_n$ that minimize (2.36), only $\mathcal{V}$ such that $D_i \in \mathcal{V}$ for all i is considered. Indeed, if D_i does not belong to $\mathcal{V}$ for some i, D_i is replaced by the point $\hat{D}_i$ which is the closest point to D_i of $\mathcal{V}$. Then, it is obvious that $|D_i, U_j| > |\hat{D}_i, U_j|$ for all j. Therefore, the value of $W(D_1, \dots, D_n)$ decreases when D_i is replaced by $\hat{D}_i$. Furthermore, the function $W(D_1, \dots, D_n)$ with $D_i \in \mathcal{V}$ for all i is a continuous function on a compact set. Hence, the minimum in the optimization problem (2.36) is achieved at some $D_1^0, \dots, D_n^0 \in \mathcal{V}$ [47]. Therefore, there exists a nonempty set of local minima, and the global minimum is achieved at one of them.

Definition 2.2 A set of points $(D_1^0, \dots, D_n^0)$ is said to be ideal if for all $i = 1, \dots, n$, $D_i^0 = C_i$, where C_i is the center of mass of the set $\mathcal{V}_i$, i.e. C_i has the coordinates (C_i^x, C_i^y), where

$$
\begin{aligned}
C_i^x &:= \frac{1}{N_i} \sum_{j \in \mathcal{V}_i} U_j^x, \\
C_i^y &:= \frac{1}{N_i} \sum_{j \in \mathcal{V}_i} U_j^y.
\end{aligned}
\tag{2.37}
$$

Here N_i is the number of users in the set $\mathcal{V}_i$, and $U_j = (U_j^x, U_j^y)$.

Proposition 2.7 A set $(D_1^0, D_2^0, \dots, D_n^0)$ is a local minimum in the optimization problem (2.36) if and only if it is ideal.

Proof. Introduce the function

$$
w_i(D_i) := \sum_{j \in \mathcal{V}_i} |D_i, U_j|^2.
\tag{2.38}
$$

Then it is obvious that

$$
w_i(D_i) = N_i(D_i^x - C_i^x)^2 + N_i(D_i^y - C_i^y)^2 - N_i C_i^{x2} - N_i C_i^{y2} + \sum_{j \in \mathcal{V}_i}(U_j^{x2} + U_j^{y2}),
\tag{2.39}
$$

where D_i^x, D_i^y are the coordinates of the UAV $i, D_i = (D_i^x, D_i^y)$. This obviously implies that if D_i^x, D_i^y are considered as variables and wish to minimize (2.38), the minimum in (2.38) is achieved at $D_i^x = C_i^x, D_i^y = C_i^y$. Moreover, consider now the straight line segment connecting the points D_i and C_i. It is obvious from (2.39) that the function (2.38) is strictly decreasing along this straight line segment from D_i to C_i. This and the obvious fact that

$$W(D_1, \ldots, D_n) = \sum_{i=1,\ldots,n} \left(w_i(D_i) + a_i^2 \right)$$

imply the statement of Proposition 2.7. □

The proposed algorithm to minimize the function (2.36) is presented in Table 2.6. As shown by Proposition 2.7, to address the problem (2.36), the area with users is divided into a set of cells, and the local optimal positions of UAVs are the centers of mass of the cells. The key is how to navigate each UAV to the centers of mass using only the information of distance. Let $D_1(t), D_2(t), \ldots, D_n(t)$ denote the UAVs locations at time $t \geq 0$. At times t_k, where $t_k > t_{k-1} > 0, k = 1, 2, 3, \ldots$, the calculated desired future locations $D_1(t_{k+1}), D_2(t_{k+1}), \ldots, D_n(t_{k+1})$ will be selected. The time interval $[t_k, t_{k+1}]$ should be large enough for the UAVs to move to the calculated desired locations. At some initial positions for the interval $[t_0, t_1]$ (see **D1** in Table 2.6), the UAVs collect the distance information to users. From t_k where $k = 1, 2, \ldots$, the proposed method navigates each UAV in a direction that leads the UAV close to the center of mass of its current cell at each step. The core method is given in **D3**. For UAV i, the center of mass $(x, y) = (C_i^x, C_i^y)$ is on the straight line defined by (2.40). How to derive (2.40) is shown in Proposition 2.8. Then, we move the UAV i to the projection of point $D_i(t_k)$ onto that straight line and such a projection is computed by (2.41). Obviously, the distance between the projection and the center of mass (C_i^x, C_i^y) is smaller than that between $D_i(t_k)$ and (C_i^x, C_i^y), i.e. the UAV i moves closer toward the center of mass.

Before proving the convergence of the proposed algorithm, an assumption is introduced. Let S_k^i denote the straight line connecting the points $D_i(t_k)$ and $D_i(\tau_k)$. The points $D_i(\tau_k)$ should be chosen to satisfy the following assumption.

Assumption 2.1 There exists a constant $\epsilon > 0$ such that the angle between the straight lines S_k^i and S_{k+1}^i is greater than ϵ for all $k = 1, 2, \ldots, i = 1, \ldots, n$.

Notice that if the trajectory of any UAV does not lie on one straight line, then $D_i(\tau_k)$ should always be chosen so that Assumption 2.1 is satisfied. In a rare case, when the trajectory of some UAV lies on a straight line, this UAV can make a manoeuvre and move along a curvy line for a short period of time. In this case, Assumption 2.1 will be satisfied. Therefore, Assumption 2.1 is not restrictive and can be easily satisfied in practice.

Table 2.6 Algorithm **D1–D4**.

D1: Select some initial UAV locations $D_1(t), \dots, D_n(t) \in \mathcal{V}$ for all $t \in [t_0, t_1]$.

For any $i = 1, 2, \dots, n$ and any $k = 1, 2, \dots$ at time t_k, do **D2–D4**.

D2: Select sometime $\tau_k^i \in [t_0, t_k)$ such that $D_i(t_k) \neq D_i(\tau_k^i)$, i.e. the UAV i was at different locations at times t_k and τ_k^i.

D3: Calculate the future UAV location $D_i(t_{k+1})$ as the projection of the point $D_i(t_k)$ onto the straight line defined by the equation

$$A_k^i x + B_k^i y = F_k^i,$$

$$A_k^i := D_i^x(\tau_k) - D_i^x(t_k),$$

$$B_k^i := D_i^y(\tau_k) - D_i^y(t_k),$$

$$F_k^i := \frac{1}{2}(D_i^{x2}(\tau_k) + D_i^{y2}(\tau_k) - D_i^{x2}(t_k) - D_i^{y2}(t_k)) + \frac{1}{2N_i} \sum_{j \in \mathcal{U}_i}(d_{ij}^2(t_k) - d_{ij}^2(\tau_k)),$$

(2.40)

where $d_{ij}(t)$ is the distance between the UAV i and the user j at time t. Hence, $D_i(t_{k+1})$ is defined as follows:

$$D_i^x(t_{k+1}) := \frac{B_k^i(B_k^i D_i^x(t_k) - A_k^i D_i^y(t_k)) + A_k^i F_k^i}{A_k^{i2} + B_k^{i2}},$$

$$D_i^y(t_{k+1}) := \frac{A_k^i(-B_k^i D_i^x(t_k) + A_k^i D_i^y(t_k)) + B_k^i F_k^i}{A_k^{i2} + B_k^{i2}}.$$

(2.41)

C4: UAV i moves from the location $D_i(t_k)$ to the location $D_i(t_{k+1})$ along the corresponding straight line.

Remark 2.4 It is clear that the proposed algorithm **(D1)–(D4)** is run by each UAV individually. At any time t_k each UAV needs to select one of its previous locations **(D2)** and then its new position at time t_{k+1} is computed by (2.41), which is very simple, and a UAV can complete the computation in $O(1)$ time. In the worst case, the complexity of each UAV in one time interval is $O(n)$, which comes from the communication phase. As aforementioned, a UAV needs to share users distance information with nearby UAVs to determine which users are covered by itself. To do this, one UAV communicate with at most $n - 1$ other UAVs. However, in practice, one UAV only needs to communicate with just several nearby UAVs. If two UAVs are far away from each other and with some other UAVs in between, they will not have overlaps in terms of user coverage. We can observe that the number of UAVs with which one UAV needs to communicate is not influenced too much by the value of n. Therefore, the proposed approach is scalable to larger n.

Now, it is the position to prove the convergence of the proposed algorithm.

Proposition 2.8 *Suppose that Assumption 2.1 holds. The UAVs trajectories constructed by the algorithm* **(D1)–(D4)** *converge to a local minimum of the optimization problem (2.36).*

Proof. It is clear that

$$\sum_{j \in \mathcal{V}_i} d_{ij}^2(t) = w_i(D_i(t)) + a_i^2 \tag{2.42}$$

for all t. Substituting $w_i(D_i(t))$ from (2.39) into (2.42), one can obtain that

$$\sum_{j \in \mathcal{V}_i} d_{ij}^2(t) = a_i^2 + N_i(D_i^x(t) - C_i^x)^2 + N_i(D_i^y(t) - C_i^y)^2 - N_i C_i^{x2} - N_i C_i^{y2}$$

$$+ \sum_{j \in \mathcal{V}_i} (U_j^{x2} + U_j^{y2}). \tag{2.43}$$

Furthermore, take (2.43) with $t = \tau_k$ and $t = t_k$ and subtract the former from the latter. Then, $x = C_i^x, y = C_i^y$ satisfy the linear equation (2.40), i.e. the center of mass point C_i belongs to the straight line defined by (2.40). Now let h_k^i denotes the distance between the center of mass C_i and the UAV position $D_i(t_k)$. Since for any k, C_i belongs to the straight line (2.40) and $D_i(t_{k+1})$ is the projection of $D_i(t_k)$ onto this line, it is obtained that $h_{k+1}^i \le h_k^i$ for all k. Therefore, h_k^i converges to some $h_0^i \ge 0$ as k tends to ∞. Now, it is the position to prove that $h_0^i = 0$. Indeed, it follows from Assumption 2.1 that for any k, the angle between two consecutive straight lines defined by (2.40) are greater than the constant $\epsilon > 0$. This implies that if $h_0^i > 0$, then

$$h_{k+1}^i - h_k^i \le -h_0^i(1 - \cos \epsilon).$$

Therefore, $h_k^i \to -\infty$ as $k \to \infty$ which is impossible because h_k^i is nonnegative. Hence, $h_0^i = 0$, and the UAVs trajectories constructed by the algorithm **(D1)–(D4)** converge to an ideal set. Therefore, it follows from Proposition 2.7 that they converge to a local minimum. This completes the proof of Proposition 2.8. □

Remark 2.5 The set of $\mathcal{V}_i$ varies with the movement of UAVs. Thus, it is possible that at some time, $\mathcal{V}_i$ becomes empty. In such a case, the UAV i hovers at the current position without moving according to Algorithm **D1–D4**.

Remark 2.6 The proposed approach does not account for the energy constraint of UAVs. To enable a long-time service, the network supplier can purchase some more UAVs for backup. These UAVs can be used when some UAVs run out of energy. Also, equipping a UAV with a solar penal enables the UAV the collect energy from the sunlight, which makes it work longer.

2.6.2 Simulation Results

In this section, computer simulation results carried out by MATLAB are presented. An area of 1 km by 1 km is considered, where some users are randomly deployed. The altitude of each UAV is fixed, and the value is between 20 and 40 m. Two main metrics are considered to evaluate the performance of our approach. The first one is the average UAV-user distance, i.e. the objective function is defined by (2.35); and the second one is the average capacity at users.

For the second metric, the genetic path loss model proposed in [41] is adopted. For the user j, the pathloss model of the LoS link is $PL = 20 \log(\frac{4\pi f_c d_j}{c}) + \eta_{LoS}$. Here, f_c is the carrier frequency, c is the light speed, and η_{LoS} is environment-dependent losses corresponding to the LoS link. Moreover, the received power at the user j is computed by $S = P_t - PL$, where P_t is the transmit power, which is identical to all the UAVs. Then, $SNR(j) = \frac{S}{N_0}$, where N_0 is the background noise. Furthermore, the capacity at user j is given by $\frac{B}{N_i} \log_2(1 + SNR(j))$, where B is the available bandwidth at UAVs and N_i is the number of users served by UAV i. The parameters take the values as follows: $\eta_{LoS} = 1$, $f_c = 1\,\text{GHz}$, $c = 3 \times 10^8\,\text{m/s}$, $P_t = 24\,\text{dBm}$, $N_0 = -104\,\text{dBm}$ and $B = 5\,\text{MHz}$.

The convergence of the proposed algorithm is demonstrated. Having 500 users randomly deployed in the considered area, 15 UAVs start from some randomly selected initial positions (see Figure 2.15a) and move to locally optimal positions (see Figure 2.15b) under the guidance of the algorithm **D1–D4**. The system converges in just 37 steps and the average UAV-user distance decreases by 31%, as shown in Figure 2.15c. To demonstrate the advantage of this algorithm, comparisons with a location-based scheme of Section 2.5 are presented in Figure 2.15c as well. Starting from the same initial positions as shown in Figure 2.15a, the compared scheme uses the users' locations directly to navigate each UAV toward the center of mass. It is clear that the performance of such a scheme depends on the percentage of available users' locations. The proposed algorithm outperforms this method in the cases with 20% and 60% available users' locations (the average UAV-user distances are 116 m and 106 m, respectively). It is also worth pointing out that the compared scheme converges faster than the proposed algorithm. The reason again lies in the knowledge of users' locations. However, as only a part of users' locations are known, the UAVs at the corresponding positions do not achieve shorter UAV-user distance than the proposed algorithm. With the increase of the percentage, the UAV-user distance achieved by the compared scheme decreases. After around 94%, the compared scheme outperforms the proposed one, see Figure 2.15d. It is easy to understand that not only the percentage of users but also the locations of users impact on the UAVs' positions. To mitigate the randomness, for a certain percentage, we independently select 100 random sets of users and use their

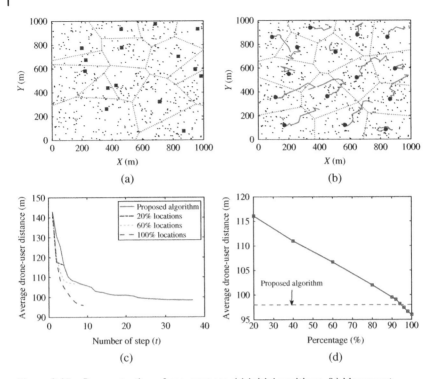

Figure 2.15 Demonstration of convergence. (a) Initial positions. (b) Movement trajectories. (c) Comparison with the benchmark scheme which assuming the availability of a certain percentage of users' locations. (d) Comparison with the benchmark scheme for different percentages of known users.

locations to compute the UAVs' locations. Thus, the results shown in Figure 2.15d are average values. When all the users' locations are known, the location-based scheme achieves lower average UAV-user distance (96 m) than the proposed one (98.4 m). Obviously, such an improvement is not so significant and very costly because it requires to collect the users' locations. Since the proposed method is location-free and achieves competitive performance, it is much more applicable in practice.

The above simulations are based on the assumption that the measured distances are accurate. However, in practice, the measured distances may contain some noise. Thus, it is important to evaluate the proposed algorithm against measurement noise. In the following simulations, we add some random noise to the measured distance. The noise takes random value between 0 and the noise level, which is between 0 and 20 m. As seen from Figure 2.16, with the increase of the noise level, the average UAV-user distance increases slightly. For the case

Figure 2.16 The influence of measurement uncertainty.

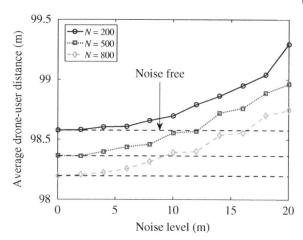

with 200 users, the average UAV-user distance is increased by about 0.8 m with noise level of 20 m, compared to the noise-free case. Another finding is that with the increase of user number, the influence of noise becomes smaller. For example, for the case with 800 users, the average UAV-user distance is only increased by 0.55 m when the noise level increases to 20 m. The reason is as follows. For a certain number of users, the distance measurement noise does influence the measurement accuracy. However, when all the users are accounted to evaluate the average UAV-user distance, the influence becomes small. Additionally, the larger the number of users, the smaller the influence of the measurement noise. Therefore, the proposed algorithm is tolerant to measurement uncertainty.

Furthermore, more simulation results for various numbers of UAVs n are presented. For each n, 100 independent simulations are conducted to mitigate the randomness of the results and all the results shown below are the average values. In Figure 2.17a, the relationship between the average UAV-user distance and n with 500 users is shown. With the increase of n, the average UAV-user distance decreases significantly. Extremely but not practically, when n increases to N, the UAV-user distance will be the smallest since each UAV can serve only one user. In Figure 2.17b, how the number of UAVs n and the number of users N influence the average user capacity is shown. In general, with the increase of n, the average capacity increases, since more bandwidth becomes available in the system. With the increase of N, the average capacity decreases, since more users share the same system bandwidth. However, the decreasing rate of the UAV-user distance and the increasing rate of average capacity both slow down with the increase of n. Thus, there is a balance between the number of UAVs and the achieved service performance. The network suppliers can choose n according to the service performance they wish to provide.

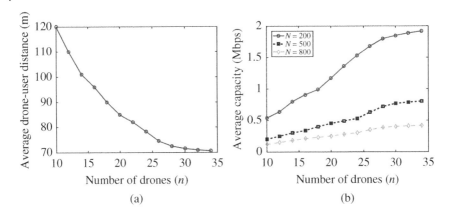

Figure 2.17 Performance for different numbers of users and numbers of UAVs. (a) Average UAV-user distance vs. the number of UAVs. (b) Average capacity vs. the number of UAVs.

2.7 Summary

In this chapter, the problems of deploying autonomous UAVs to provide cellular service to UEs is considered. Two typical scenarios are studied. The first scenario is in urban areas, and the UAVs are deployed over streets to avoid collision with buildings. The second scenario is in disaster areas. We formulate several optimization problems to optimize the quality of service provided by the UAVs, and decentralized algorithms are presented to address these problems. Extensive computer simulations were conducted and the results demonstrate the effectiveness of the proposed methods. In Chapter 3, we will discuss the aerial surveillance problems. This kind of problems shares some similarities with the problems considered in this chapter, but there are also some difference between the underlying coverage models. We will provide more discussion on it in Chapter 3.

References

1 T. Nakamura, S. Nagata, A. Benjebbour, Y. Kishiyama, T. Hai, S. Xiaodong, Y. Ning, and L. Nan, "Trends in small cell enhancements in LTE advanced," *IEEE Communications Magazine*, vol. 51, no. 2, pp. 98–105, 2013.

2 M. Ding, P. Wang, D. López-Pérez, G. Mao, and Z. Lin, "Performance impact of LOS and NLOS transmissions in dense cellular networks," *IEEE Transactions on Wireless Communications*, vol. 15, no. 3, pp. 2365–2380, 2016.

3 A. Al-Hourani, S. Kandeepan, and A. Jamalipour, "Modeling air-to-ground path loss for low altitude platforms in urban environments," in *IEEE Global*

Communications Conference (GLOBECOM), (Austin, TX, USA), pp. 2898–2904, Dec. 2014.

4 M. Mozaffari, W. Saad, M. Bennis, and M. Debbah, "Drone small cells in the clouds: design, deployment and performance analysis," in *Proceedings IEEE Global Communications Conference (GLOBECOM)*, (San Diego, CA, USA), pp. 1–6, Dec. 2015.

5 R. I. Bor-Yaliniz, A. El-Keyi, and H. Yanikomeroglu, "Efficient 3-D placement of an aerial base station in next generation cellular networks," in *Proceedings of IEEE International Conference Communications (ICC)*, (Kuala Lumpur, Malaysia), pp. 1–5, May 2016.

6 M. Alzenad, A. El-Keyi, F. Lagum, and H. Yanikomeroglu, "3D placement of an unmanned aerial vehicle base station (UAV-BS) for energy-efficient maximal coverage," *IEEE Wireless Communications Letters*, vol. 6, no. 4, pp. 434–437, 2017.

7 V. Sharma, M. Bennis, and R. Kumar, "UAV-assisted heterogeneous networks for capacity enhancement," *IEEE Communications Letters*, vol. 20, no. 6, pp. 1207–1210, 2016.

8 E. Kalantari, H. Yanikomeroglu, and A. Yongacoglu, "On the number and 3D placement of drone base stations in wireless cellular networks," in *Proceedings of IEEE Vehicular Technology Conference*, (Montréal, QC, Canada), pp. 1–6, 2016.

9 R. K. Ganti, F. Ye, and H. Lei, "Mobile crowdsensing: current state and future challenges," *IEEE Communications Magazine*, vol. 49, no. 11, pp. 32–39, 2011.

10 T. Chen, M. A. Kaafar, and R. Boreli, "The where and when of finding new friends: analysis of a location-based social discovery network.," in *Proceedings of the 7th International Conference Weblogs Soc Media (ICWSM)*, pp. 61–70, 2013.

11 T. M. Cheng and A. V. Savkin, "Decentralized control of mobile sensor networks for asymptotically optimal blanket coverage between two boundaries," *IEEE Transactions on Industrial Informatics*, vol. 9, no. 1, pp. 365–376, 2013.

12 B. Galkin, J. Kibilda, and L. A. DaSilva, "Deployment of UAV-mounted access points according to spatial user locations in two-tier cellular networks," in *Wireless Days (WD)*, pp. 1–6, 2016.

13 J. Lyu, Y. Zeng, R. Zhang, and T. J. Lim, "Placement optimization of UAV-mounted mobile base stations," *IEEE Communications Letters*, vol. 21, pp. 604–607, 2017.

14 H. Huang, A. V. Savkin, M. Ding, and M. A. Kaafar, "Optimized deployment of drone base station to improve user experience in cellular networks," *Journal of Network and Computer Applications*, vol. 144, pp. 49–58, 2019.

15 "2008 Sichuan earthquake," Accessed on 20 July 2019. Online: https://en .wikipedia.org/wiki/2008_Sichuan_earthquake.

16 A. V. Savkin and H. Huang, "A method for optimized deployment of a network of surveillance aerial drones," *IEEE Systems Journal*, vol. 13, no. 4, pp. 4474–4477, 2019.

17 A. V. Savkin and H. Huang, "Asymptotically optimal deployment of drones for surveillance and monitoring," *Sensors*, vol. 19, no. 9, p. 2068, 2019.

18 A. Fotouhi, M. Ding, and M. Hassan, "Flying drone base stations for macro hotspots," *IEEE Access*, vol. 6, pp. 19530–19539, 2018.

19 E. Kalantari, H. Yanikomeroglu, and A. Yongacoglu, "On the number and 3D placement of drone base stations in wireless cellular networks," in *IEEE Vehicular Technology Conference*, pp. 1–6, Sept. 2016.

20 H. Huang and A. V. Savkin, "An algorithm of efficient proactive placement of autonomous drones for maximum coverage in cellular networks," *IEEE Wireless Communications Letters*, vol. 7, no. 6, pp. 994–997, 2018.

21 H. Huang and A. V. Savkin, "A method for optimized deployment of unmanned aerial vehicles for maximum coverage and minimum interference in cellular networks," *IEEE Transactions on Industrial Informatics*, vol. 15, no. 5, pp. 2638–2647, 2019.

22 A. V. Savkin and H. Huang, "Deployment of unmanned aerial vehicle base stations for optimal quality of coverage," *IEEE Wireless Communications Letters*, vol. 8, no. 1, pp. 321–324, 2019.

23 A. V. Savkin and H. Huang, "Range-based reactive deployment of autonomous drones for optimal coverage in disaster areas," *IEEE Transactions on Systems, Man, and Cybernetics: Systems*, vol. 51, no. 7, pp. 4606–4610, 2021.

24 V. Sharma, K. Srinivasan, H.-C. Chao, K.-L. Hua, and W.-H. Cheng, "Intelligent deployment of UAVs in 5G heterogeneous communication environment for improved coverage," *Journal of Network and Computer Applications*, vol. 85, pp. 94–105, 2017.

25 I. Bor-Yaliniz, S. S. Szyszkowicz, and H. Yanikomeroglu, "Environment-aware drone-base-station placements in modern metropolitans," *IEEE Wireless Communications Letters*, vol. 7, no. 3, pp. 372–375, 2017.

26 P. Yang, X. Cao, C. Yin, Z. Xiao, X. Xi, and D. Wu, "Proactive drone-cell deployment: overload relief for a cellular network under flash crowd traffic," *IEEE Transactions on Intelligent Transportation Systems*, vol. 18, no. 10, pp. 2877–2892, 2017.

27 Z. Becvar, M. Vondra, P. Mach, J. Plachy, and D. Gesbert, "Performance of mobile networks with UAVs: can flying base stations substitute ultra-dense small cells?," in *the 23th European Wireless Conference*, pp. 1–7, VDE, 2017.

28 M. Chen, M. Mozaffari, W. Saad, C. Yin, M. Debbah, and C. S. Hong, "Caching in the sky: proactive deployment of cache-enabled unmanned aerial vehicles for optimized quality-of-experience," *IEEE Journal on Selected Areas in Communications*, vol. 35, no. 5, pp. 1046–1061, 2017.

29 Y. Zeng, R. Zhang, and T. J. Lim, "Wireless communications with unmanned aerial vehicles: opportunities and challenges," *IEEE Communications Magazine*, vol. 54, pp. 36–42, 2016.

30 R. Waterhouse and D. Novack, "Realizing 5G: microwave photonics for 5G mobile wireless systems," *IEEE Microwave Magazine*, vol. 16, pp. 84–92, 2015.

31 S. Sekander, H. Tabassum, and E. Hossain, "Multi-tier drone architecture for 5G/B5G cellular networks: challenges, trends, and prospects," *IEEE Communications Magazine*, vol. 56, pp. 96–103, 2018.

32 H. Ghazzai, E. Yaacoub, M.-S. Alouini, Z. Dawy, and A. Abu-Dayya, "Optimized LTE cell planning with varying spatial and temporal user densities," *IEEE Transactions on Vehicular Technology*, vol. 65, no. 3, pp. 1575–1589, 2016.

33 W. Zhao, S. Wang, C. Wang, and X. Wu, "Approximation algorithms for cell planning in heterogeneous networks," *IEEE Transactions on Vehicular Technology*, vol. 66, no. 2, pp. 1561–1572, 2017.

34 A. Fotouhi, M. Ding, and M. Hassan, "Dynamic base station repositioning to improve spectral efficiency of drone small cells," in *IEEE WoWMoM Workshop on Internet of Things-Smart Objects and Services (IoT-SoS)*, 2017.

35 L. Wang, B. Hu, and S. Chen, "Energy efficient placement of a drone base station for minimum required transmit power," *IEEE Wireless Communications Letters*, vol. 9, no. 12, pp. 2010–2014, 2020.

36 D. Takaishi, H. Nishiyama, N. Kato, and R. Miura, "A dynamic trajectory control algorithm for improving the probability of end-to-end link connection in unmanned aerial vehicle networks," in *International Conference on Personal Satellite Services*, pp. 94–105, Springer, 2016.

37 X. Li, D. Guo, H. Yin, and G. Wei, "Drone-assisted public safety wireless broadband network," in *IEEE Wireless Communications and Networking Conference Workshops (WCNCW)*, pp. 323–328, Mar. 2015.

38 D. G. Cileo, N. Sharma, and M. Magarini, "Coverage, capacity and interference analysis for an aerial base station in different environments," in *International Symposium on Wireless Communication Systems (ISWCS)*, (Bologna, Italy), pp. 281–286, Aug. 2017.

39 F. Lagum, I. Bor-Yaliniz, and H. Yanikomeroglu, "Strategic densification with UAV-BSs in cellular networks," *IEEE Wireless Communications Letters*, vol. 7, pp. 384–387, 2018.

40 C. Pan, H. Ren, Y. Deng, M. Elkashlan, and A. Nallanathan, "Joint blocklength and location optimization for URLLC-enabled UAV relay systems," *IEEE Communications Letters*, vol. 23, pp. 498–501, 2019.

41 A. Al-Hourani, S. Kandeepan, and S. Lardner, "Optimal LAP altitude for maximum coverage," *IEEE Wireless Communications Letters*, vol. 3, pp. 569–572, 2014.

42 Z. Yang, C. Pan, M. Shikh-Bahaei, W. Xu, M. Chen, M. Elkashlan, and A. Nallanathan, "Joint altitude, beamwidth, location, and bandwidth optimization for UAV-enabled communications," *IEEE Communications Letters*, vol. 22, pp. 1716–1719, 2018.

43 H. Huang and A. V. Savkin, "Reactive 3D deployment of a flying robotic network for surveillance of mobile targets," *Computer Networks*, vol. 161, pp. 172–182, 2019.

44 M. Mozaffari, W. Saad, M. Bennis, and M. Debbah, "Mobile unmanned aerial vehicles (UAVs) for energy-efficient internet of things communications," *IEEE Transactions on Wireless Communications*, vol. 16, no. 11, pp. 7574–7589, 2017.

45 M. Mozaffari, A. T. Z. Kasgari, W. Saad, M. Bennis, and M. Debbah, "Beyond 5G with UAVs: foundations of a 3D wireless cellular network," *IEEE Transactions on Wireless Communications*, vol. 18, no. 1, pp. 357–372, 2018.

46 H. Huang and A. V. Savkin, "An algorithm of reactive collision free 3D deployment of networked unmanned aerial vehicles for surveillance and monitoring," *IEEE Transactions on Industrial Informatics*, vol. 16, no. 1, pp. 132–140, 2020.

47 A. N. Kolmogorov and S. V. Fomin, *Introductory real analysis*. New York: Dover, 1975.

48 A. Al-Hourani, S. Kandeepan, and S. Lardner, "Optimal LAP altitude for maximum coverage," *IEEE Wireless Communications Letters*, vol. 3, no. 6, pp. 569–572, 2014.

49 A. Fotouhi, M. Ding, and M. Hassan, "Understanding autonomous drone maneuverability for internet of things applications," in *the 18th International Symposium on A World of Wireless, Mobile and Multimedia Networks (WoW-MoM)*, pp. 1–6, IEEE, 2017.

50 "3GPP TR 36.828: Further enhancements to LTE time division duplex (TDD) for (DL-UL) interference management and traffic adaptation," 2012.

51 I. R. Petersen and A. V. Savkin, *Robust Kalman filtering for signals and systems with large uncertainties*. Birkhauser, Boston, MA, 1999.

52 P. N. Pathirana, A. V. Savkin, and S. Jha, "Location estimation and trajectory prediction for cellular networks with mobile base stations," *IEEE Transactions on Vehicular Technology*, vol. 53, pp. 1903–1913, 2004.

53 P. N. Pathirana, N. Bulusu, A. V. Savkin, and S. Jha, "Node localization using mobile robots in delay-tolerant sensor networks," *IEEE Transactions on Mobile Computing*, vol. 4, no. 3, pp. 285–296, 2005.

54 D. Bailey, M. Trotter, C. Knight, and M. Thomas, "Use of GPS tracking collars and accelerometers for rangeland livestock production research," *Journal of Animal Science*, vol. 95, p. 360, 2017.

55 D. A. McGranahan, B. Geaumont, and J. W. Spiess, "Assessment of a livestock GPS collar based on an open-source datalogger informs best practices for logging intensity," *Ecology and Evolution*, vol. 8, no. 11, pp. 5649–5660, 2018.

3

Deployment of UAVs for Surveillance of Ground Areas and Targets

3.1 Introduction

Among the various promising applications of unmanned aerial vehicles (UAVs), aerial surveillance is the one that has attracted great attention of practitioners in recent years, which can be used for the protection of assets, people or objects, the investigation of crimes, and intelligence gathering. A main advantage of UAVs for surveillance is the high probability of having line-of-sight (LoS) with ground objects, which may be hardly achieved by ground-based sensing units. Under the topic of aerial surveillance, there are several interesting topics having a significant impact on the quality of surveillance. The first one is the video- and image-processing techniques [1–3]. Such techniques are the fundamental tools for UAVs to detect and track targets of interest and extract certain features of targets, without which it will be tedious for human operators to conduct the target detection and tracking. The size of the target object in the video image may be as large as the background noise, and the onboard camera may also experience rapid oscillations and translations which further prevent the manual target detection [4]. The second topic relevant to aerial surveillance by UAVs is the deployment problem. This problem generally focuses on finding the minimum number of UAVs to achieve a given request such as the full coverage of a target area [5, 6] or where to deploy a given number of UAVs to achieve the best quality of surveillance [7, 8]. It is worth noting that from the point of coverage control, the wireless coverage problems considered in Chapter 2 share some similarities with the aerial surveillance problems considered in this chapter. The main difference lies in the coverage model. For surveillance, the coverage mainly depends on the field of view (FoV) and LoS, while for wireless communications, the coverage is dependent on some communication metrics such as signal-to-noise-ratio (SNR) and receiving rate.

In this chapter, we pay attention to the deployment of UAVs. We discuss two important approaches developed recently for area surveillance and target

Autonomous Navigation and Deployment of UAVs for Communication, Surveillance and Delivery,
First Edition. Hailong Huang, Andrey V. Savkin, and Chao Huang.

surveillance, respectively. For the former, we focus on finding the minimum number of UAVs and the positions for them to fully cover a given area of interest. For the latter, we focus on the optimal deployment of a given number of UAVs to maximize certain metrics of quality of surveillance for some ground target objects. More specifically, we will cover several typical and practical cases. For the area surveillance, we will discuss the coverage of a planer area (and/or with holes representing no-fly zones or the like). We will also consider scenarios with lots of obstacles that are potentially block of LoS between UAVs and objects. For the ground object surveillance, we will extensively discuss the optimal deployment problem for 2D and 3D cases, where the UAVs remain their altitudes and can change the altitudes, respectively.

In surveillance applications, a typical situation is that a fleet of UAVs, carrying some specific sensors such as ground-facing cameras, monitors, and some ground objects. The cameras can see a disc on the ground and the altitude of the UAV as well as the visibility angle have significant impacts on this visible region. In this kind of applications, a significant technical problem is to deploy the minimal number of UAVs to cover a given ground area completely. We present a method of deploying a number of UAVs so that every point of a given ground region is seen from at least one UAV. Moreover, we prove that the proposed deployment is asymptotically optimal in the sense that the ratio of the number of UAVs deployed and the minimal possible number of UAVs needed to cover the ground region converges to one as the area of the ground region tends to infinity. This theoretical result is derived from so-called "Kershner's theorem" [9], a powerful and elegant tool of combinatorial geometry. Furthermore, the proposed algorithm is based on constructing a triangulation of the ground plane consisting of congruent equilateral triangles, and not all but most of the UAVs are deployed at vertices of the triangulation. This makes the proposed method easily understandable and very computationally efficient. Moreover, the proposed algorithm results in a deployment in which positions of UAVs are separated from each other which greatly reduces the danger of collisions.

For the case with obstacles such as buildings in the areas, a stronger require-ment of monitoring applications is to ensure that any point of the area is seen from some UAV, and a very difficult and practically important extension of this problem is deployment of UAVs over very uneven terrains, rather than flat areas. Uneven terrains are geometrically complex environments that cannot be approx-imated with a sufficient accuracy by a plane. In such environments, the visibility cones of UAVs' cameras may be occluded by buildings, walls, mountains, hills, etc. Very uneven terrains are quite typical for dense urban areas with tall buildings and narrow streets [10, 11]. This problem will get especially important in the future with the use of small low-flying UAVs [10]. To solve this problem, a constructive solution of this UAV deployment problem that may be called a UAV version of the

3D Art Gallery Problem (AGP) is presented, and we use the three-coloring method presented in [12] to find the positions for UAVs.

For the target surveillance when UAVs are deployed at a given altitude, we consider the problem of maximizing coverage by a set of UAVs, which should form a connected graph with some fixed ground nodes (GNs). The captured information is delivered by UAVs to a central unit via GNs. A distributed coverage maximizing algorithm to find locally optimal positions of UAVs is proposed. It requires only local information rather than global information; thus, it is very easily implementable. In particular, each UAV should only know some precomputed probability density within its visibility cone and the positions of nearby UAVs. The presented algorithm is decentralized and scalable to a large number of UAVs. It is proved that the algorithm converges to a local maximum in a finite number of steps.

When UAVs are allowed to adjust their altitudes, a novel coverage model characterizing the quality of coverage (QoC) of a target by a UAV is discussed. Beyond the basic disk model, this model distinguishes targets at different positions inside the cone. Based on this model, a reactive algorithm is developed to deploy UAVs in continuous 3D space. For a given fleet of UAVs, the objective is to maximize the overall QoC of a given set of targets. It is assumed that each UAV can estimate the locations of targets which are seen by the on-board camera, using some available image-processing techniques. Also, each UAV can measure the positions of other nearby UAVs. With these two pieces of information as input, each UAV decides its movement individually. The movement of each UAV is separated into the horizontal submovement and the vertical submovement. The presented algorithm consists of two navigation laws for these two submovements, respectively. Several constraints have to be satisfied when UAVs move. There is a set of GNs that need to collect some required information from UAVs timely. The UAVs are required to maintain valid links with GNs all the time, so that any required information can be delivered to the data centers without significant delays. Also, some pairs of UAVs are required to always keep communication links between each other. Furthermore, the UAVs need to avoid collisions with each other [13], and they are not allowed to enter some no-flight areas. The main results of this chapter were originally published in [5–8].

The remainder of this chapter is organized as follows: Section 3.2 discusses some closely related references. Section 3.3 presents a method for the surveillance of a flat ground area using the minimum number of UAVs, and Section 3.4 develops an algorithm of deployment of UAVs for surveillance of uneven ground areas. Section 3.5 discusses a target surveillance problem when the UAVs are deployed at a fixed altitude, and Section 3.6 considers a scenario where ground-monitoring UAVs can adjust their altitudes. Finally, Section 3.7 summarizes this chapter and points out several promising directions for future research.

3.2 Related Work

In surveillance and monitoring applications, UAVs are mostly equipped with cameras. They fly into the sky and monitor ground targets of interest, such as humans, buildings, pipelines, roads, vehicles, landmarks and animals, for the purpose of surveillance and security, etc. [14–17]. A common scenario is when each UAV is equipped with a ground-facing camera with some visibility angle. The camera can see a circular area on the ground, and the radius of this area depends on the altitude and the visibility angle. If a point is within this circle, it is considered to be covered by the UAV. The altitude of each UAV must be in a given range. A well-studied scenario is using a fleet of UAVs to monitor a given ground area, and a general approach is to partition the area first and then plan paths for UAVs in the subareas [18]. A shortcoming of this partition-based method is that it does not guarantee the coverage of any point of the area at any time. Once the motion pattern of the UAVs is learned, intruders may be able to avoid the monitoring. This chapter [16] proposes a decoupling method to find the minimum number and the UAVs' positions. The paper [19] focuses on continuously covering a set of mobile targets considering the constraint of energy capacity of UAVs. The review [20] proposes a localized heuristic algorithm for a similar problem with [19]. Moreover, the publication [21] examines the case with multiple subareas accounting the charging requirement of UAVs. The common feature of these references is that they develop centralized deployment algorithms that are based on either grids or a given discrete candidate set. Obviously, solutions that are optimal in some discrete set are not guaranteed to be optimal in continuous regions.

Uneven terrains are geometrically complex environments that cannot be approximated with a sufficient accuracy by a plane. In such environments, the visibility cones of UAVs' cameras may be occluded by buildings, walls, mountains, hills, etc. Very uneven terrains are quite typical for dense urban areas with tall buildings and narrow streets. One approach to this problem is to first find a set of positions such that every point on the terrain can be seen from at least one of the positions in the set [22, 23] and then construct a tour for the UAV to visit these positions such that every position is visited exactly once and the tour completion time is minimized.

The AGP is a well-known problem of combinatorial/computational geometry that deals with determining the minimal number of observers necessary to cover an art gallery room such that every point is seen by at least one observer. This problem was formulated by in [24] and is well studied, especially for the 2D case [24–26]. In [27], the Art Gallery Theorem was published that gives an upper bound on the minimal number of observers in the 2D AGP. The proof of [27] was later simplified in [12] via a so-called "three-coloring method." In [28], a 3D version of the AGP was considered where observers are to be placed on uneven terrain modeled as a nonconvex polytope. In this chapter, a UAV version of the 3D AGP in which observers (UAVs) are to be placed above uneven terrain in a certain range of altitudes (not on the terrain as in [28]) is considered.

In the surveillance of targets rather than the whole area, the QoC is a key system metric. Several interesting problems have been studied. To track a single target, Ref. [29] proposes a simple control strategy to reach a given target view angle; and Ref. [30] focuses on optimal routing for two UAVs cooperatively tracking a moving target. For multiple targets, in [19], the problem of minimizing the number of UAVs to fully cover a given set of targets is considered. In [14], the authors study the joint problem of coverage, connectivity, and charging UAVs. In [15], the authors address the optimization problem of covering a set of targets with a fleet of UAVs. The goal is to deploy a connected set of UAVs continuously monitoring the targets, and collecting or sending data to them. As an extension, the authors of [16] consider the problem of monitoring a set of mobile sensors. The authors of [31] provide a power efficient and reliable scheduling by adjusting the UAVs' positions to ensure the surveillance of all the targets. With the objective of maximizing the number of covered targets, the authors of [32] propose a reactive algorithm to navigate each flying robot based on virtual forces including hotspots attractive force, target attractive force, nearby robot repulsive force, and obstacle repulsive force. It is [33] proposed that a scheme to cover all ground users using a minimum number of UAVs be applied.

3.3 Asymptotically Optimal UAV Deployment for Surveillance of a Flat Ground Area

We start from a simple scenario where the area of interest is flat, and we consider a static deployment problem for the purpose of full coverage of the area of interest. In Section 3.3.1, we present the symbols first and then state the problem of interest. The deployment algorithm is presented in Section 3.3.2.

3.3.1 Problem Statement

Let (x, y) be Cartesian coordinates on the ground plane and z be the coordinate axis perpendicular to the ground plane, where the equation $z = 0$ describes the ground plane itself. Moreover, let D be a given bounded and Lebesgue measurable region [34] of the ground plane $z = 0$ with piecewise smooth boundary. Our objective is to deploy a number of UAVs to monitor the corresponding area of the ground region D. Also, let Z_{min} and Z_{max} be given minimum and maximum altitudes for

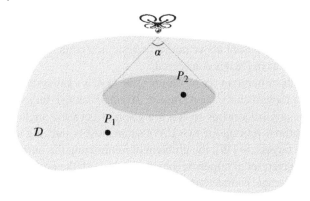

Figure 3.1 The visibility cone. Point P_2 can be seen by the UAV, while point P_1 cannot.

UAV deployment, $Z_{max} > Z_{min} > 0$. Assume that each UAV can be deployed only at some points (x_u, y_u, z_u) such that

$$(x_u, y_u) \in D, \quad z_u \in [Z_{min}, Z_{max}]. \tag{3.1}$$

Moreover, the UAVs have a given observation angle $0 < \alpha < \pi$, which defines the visibility cone of each UAV, so that a UAV with the coordinates (x_u, y_u, z_u) can only see points (x, y, z) of the ground that are inside of the circle of radius

$$r := \tan\left(\frac{\alpha}{2}\right) z_u \tag{3.2}$$

centered at (x_u, y_u). A point P on the terrain is visible from a UAV if P is inside of the visibility cone of the UAV, see Figure 3.1.

Definition 3.1 Deployment of a number of UAVs is said to be *covering* if the constraints (3.1) are satisfied and every point on the ground region D can be seen by at least one of the UAVs.

It is obvious that when we increase the altitude of a UAV, its visibility cone is increasing. Therefore, if there exists a covering deployment at some altitude z_u, then there exists a covering deployment at any altitude z such that $Z_{max} \geq z > z_u$. In this chapter, the goal is to place the UAVs at the lowest possible altitude to reduce the distances between the UAVs and the points of the monitored ground area.

Problem Statement: Let N be the number of UAVs, and z_u be a given UAVs' altitude satisfying (3.1). Our aim is to construct, if possible, a covering deployment, i.e. to deploy N UAVs at the altitude z_u so that any point of the ground region D is seen by at least one UAV. Moreover, we want to find the minimal altitude z_u for which there still exists a covering deployment with N UAVs.

Remark 3.1 Many publications consider various problems of path planning for UAVs where the objective is to get information about some area of interest, see, e.g. [35, 36]. The difference with the problem studied in this chapter is that we consider a problem of deploying UAVs in steady positions to achieve constant surveillance of each point of an area of interest.

Remark 3.2 A very effective tool for many UAV placement problems is convex programming, see e.g. [37]. In this chapter, we consider the problem of finding the minimal number of UAVs that cover some given bounded ground area. The conic visibility sector of each UAV is convex, indeed. However, the union of visibility sectors of several UAVs is nonconvex. Furthermore, the ground area to be covered may be nonconvex as well. Therefore, the problem under consideration is nonconvex.

3.3.2 Deployment Algorithm

On the plane $z = z_u$ that is parallel to the ground, we will consider triangulations consisting of equilateral triangles with the side $r_t = \sqrt{3}r$ where r is the radius defined by (3.2), see Figure 3.2. Any such triangulation $\mathcal{T}(\lambda, x_0, y_0)$ is defined by the angle $\lambda \in [0, \frac{\pi}{3})$, and the parameters x_0, y_0 that are the Cartesian coordinates of some point inside a rhombus with the side r_t consisting of two equilateral triangles. Here λ is the angle between the coordinate axis x and one of the directions of the triangulation $\mathcal{T}(\lambda, x_0, y_0)$, and x_0, y_0 are the coordinates of one of the vertices of this triangulation, see Figure 3.2. Moreover, for an equilateral triangle, we will consider its center C and three congruent Voronoi cells consisting of points of the triangle for which a certain vertex of the triangle is closer than the other two vertices [38], see Figure 3.3.

Figure 3.2 A triangulation $\mathcal{T}(\lambda, x_0, y_0)$ consisting of equilateral triangles.

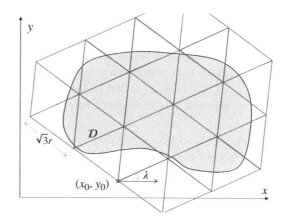

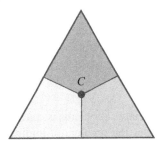

Figure 3.3 The center of an equilateral triangle and the three congruent Voronoi cells which are in different colors.

The proposed deployment algorithm of deployment UAVs on the plane $z = z_u$ consists of the following steps:

Step A1: We choose some λ, x_0, y_0 and construct the triangulation $\mathcal{T}(\lambda, x_0, y_0)$.

Step A2: We deploy UAVs at all vertices of the triangulation $\mathcal{T}(\lambda, x_0, y_0)$ that belong to the ground region $\mathcal{D}$.

Step A3: We consider any triangle $\mathcal{W}$ of the triangulation $\mathcal{T}(\lambda, x_0, y_0)$ such that it has some overlap with the ground region $\mathcal{D}$, but not all the three vertices of the triangle are inside $\mathcal{D}$, i.e. some of its vertices were not occupied by UAVs in Step **A2**. Let l be the number of such unoccupied vertices of $\mathcal{W}, l \in \{1, 2, 3\}$.

Step A4: Consider the Voronoi cell of $\mathcal{W}$ corresponding to some unoccupied vertex of $\mathcal{W}$. If this Voronoi cell contains some point of $\mathcal{D}$ that is not covered by any UAV deployed in Step **A2**, we deploy a UAV at the point that is closest to the triangle center C among all the points of this Voronoi cell belonging to $\mathcal{D}$.

Step A5: If $l \geq 2$, we consider the Voronoi cell of $\mathcal{W}$ corresponding to some unoccupied vertex of $\mathcal{W}$ other than considered in **A4**. If this Voronoi cell contains some point of $\mathcal{D}$ that is not covered by any UAV deployed in Steps **A2, A4**, we deploy a UAV at the point that is closest to the triangle center C among all the points of this Voronoi cell belonging to $\mathcal{D}$.

Step A6: If $l = 3$, we consider the Voronoi cell of $\mathcal{W}$ corresponding to the third unoccupied vertex of $\mathcal{W}$. If this Voronoi cell contains some point of $\mathcal{D}$ that is not covered by any UAV deployed in Steps **A2, A4, A5**, we deploy a UAV at the point that is closest to the triangle center C among all the points of this Voronoi cell belonging to $\mathcal{D}$.

Remark 3.3 It is obvious that for any triangle from **A3**, we deploy no more than l new UAVs at its vertices.

Proposition 3.1 *Let an altitude z_u be given, $Z_{min} \leq z_u \leq Z_{max}$. Then the deployment algorithm **A1**–**A6** is covering.*

Proof. It is obvious that the distance between any two points in a certain Voronoi cell of a certain equilateral triangle with the side length $\sqrt{3}r$ is no greater than r, see Figure 3.3. Hence, if for any point of the ground region D, there exists a UAV located at some point of the same Voronoi cell of our triangulation to which this point of D belongs to, then point is covered by this UAV. It is obvious that the deployment algorithm **A1–A6** is placing a UAV at some point of any Voronoi cell that contains at least one point of the region D. This completes the proof of Proposition 3.1. □

We will also analyze the optimality of the algorithm **A1–A6**. Let $\gamma > 1$ be some given number. Introduce the region D_γ obtained from the region D by the linear transformation that maps any point $(x, y) \in \mathbf{D}$ to the point $(\gamma x, \gamma y)$, see Figure 3.4. So the region D_γ is similar to D but larger. It is also obvious that

$$A(D_\gamma) = \gamma^2 A(D), \tag{3.3}$$

where $A(\cdot)$ denotes the area of a planar region.

Furthermore, consider some family $\mathcal{F}(\gamma)$ of covering deployments of the regions D_γ for all $\gamma > 1$. Let $N(\gamma)$ be the number of UAVs in the deployment $\mathcal{F}(\gamma)$. Let $M(\gamma)$ be the minimal possible number of UAVs in all covering deployments of the region D_γ. It is clear that since the region D_γ increase as γ is increases, both $M(\gamma)$ and $N(\gamma)$ tend to infinity as γ tends to infinity.

Definition 3.2 A family $\mathcal{F}(\gamma)$ of covering deployments of the regions D_γ is said to be asymptotically optimal, if

$$\lim_{\gamma \to \infty} \frac{N(\gamma)}{M(\gamma)} = 1. \tag{3.4}$$

In other words, a covering deployment is asymptotically optimal, in the sense that as the ground region becomes larger, the number of UAVs in this deployment

Figure 3.4 Constructing region D_γ from region D.

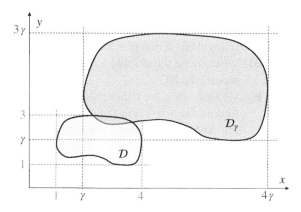

becomes close to the minimal possible number of UAVs for any covering deployment.

Proposition 3.2 *Let an altitude z_u be given, $Z_{min} \leq z_u \leq Z_{max}$. Let $\mathcal{F}(\gamma)$ be the family of covering deployments of the regions $\mathcal{D}_\gamma$ constructed by the deployment algorithm **A1–A6**. Then $\mathcal{F}(\gamma)$ is asymptotically optimal. Moreover,*

$$\lim_{\gamma \to \infty} \frac{N(\gamma)}{\gamma^2} = \frac{2A(\mathcal{D})}{3\sqrt{3}\tan^2(\frac{\alpha}{2})z_u^2}. \tag{3.5}$$

Proof. Indeed, let $\hat{N}(\gamma)$ be the number of vertices of the triangulation $\mathcal{T}(\lambda, x_0, y_0)$ constructed in Step **A1** such that there is at least one triangle of this triangulation with its vertex intersecting with the region $\mathcal{D}_\gamma$. Then it is obvious that

$$\lim_{\gamma \to \infty} \frac{N(\gamma)}{\hat{N}(\gamma)} = 1. \tag{3.6}$$

Furthermore, covering of a ground region by minimal number of UAVs deployed at the altitude z_u is equivalent to covering of minimal number of discs of radius r. Introduce the variable $\epsilon := r\gamma^{-1}$. It is obvious that $\epsilon \to 0$ as $\gamma \to \infty$. Therefore, the problem of covering the region $\mathcal{D}_\gamma$ by minimal number of discs of radius r as $\gamma \to \infty$ is equivalent to the problem of covering the region $\mathcal{D}$ by minimal number of discs of radius ϵ as $\epsilon \to 0$. Therefore, Eq. (3.5) and Kershner's theorem [9] imply the statement of this proposition. This completes the proof of Proposition 3.2. □

Let N be the number of UAVs we can deploy. Now, we present our deployment algorithm that includes a search for the minimal altitude z_u.

Step B1: We start with the altitude $z_u := Z_{max}$. We make some search in the space of parameters (λ, x_0, y_0), where the angle $\lambda \in [0, \frac{\pi}{3})$, and the parameters x_0, y_0 that are the Cartesian coordinates of some point inside a rhombus with the side r_t consisting of two equilateral triangles, and apply the deployment algorithm **A1–A6** to the triangulation $\mathcal{T}(\lambda, x_0, y_0)$. We take triangulations that give a minimal number of UAVs.

Step B2: We decrease the altitude z_u by some small value $\epsilon > 0$: $z_u := z_u - \epsilon$ and repeat step **B1**.

Step B3: We stop when we either reach the minimum altitude Z_{min} or failed to find a triangulation at the current altitude z_u for which the algorithm **A1–A6** requires no more than N UAVs.

Remark 3.4 In **B1**, we conduct a complete search for all the feasible configurations of (λ, x_0, y_0). As the triangles in the triangulation are all similar, we can discretize the area into grids with a given resolution and only consider the grids falling into one triangle. At each grid, the angle λ can take value in the range $[0, \frac{\pi}{3})$. Setting another resolution for λ, a full set of configurations of (λ, x_0, y_0) is obtained. Given the triangle side r_t, the number of the initial configurations depends on the selection of the two resolutions.

Remark 3.5 The algorithm **B1–B3** is a recursive algorithm. It will check a lower altitude if the current altitude enables the full coverage with N UAVs. In the worst case, this algorithm repeats for $\lceil \frac{Z_{max}-Z_{min}}{\epsilon} \rceil$ times.

Remark 3.6 It should be pointed out that the proposed algorithms **A1–A6** and **B1–B3** result in deployments in which not all but most UAVs are placed at vertices of a triangulation consisting of congruent equilateral triangles. This means that the positions of UAVs are usually sufficiently separated from each other. Such a separation prevents collisions between UAVs, which is crucial for real-life UAV systems, see, e.g. [39–41].

Remark 3.7 A very practically important extension of the obtained results would be the case of complex ground structures, e.g. environments with high-rise buildings causing blocking of view from some UAVs to certain fragments of the ground area that are inside of UAVs' visibility cones. Some version of such an extension will be presented in Section 3.4 of this chapter, see also [6].

Remark 3.8 The reason why we need to place UAVs in a triangular formation is as follows. We state our problem as finding the minimal number of UAVs that cover each point of a bounded ground region of interest. Proposition 3.2 shows that the optimal solution of this problem is a triangular formation. More precisely, it is an almost triangular formation as not all but most of UAVs need to be placed at vertices of a triangular grid. Of course, there are many other problem statements which lead to other deployment patterns. Those problem statements may have their advantages in certain real-life problems. However, in our problem statement, a triangular deployment is the optimal solution which has been proved in a mathematically rigorous way in Proposition 3.2.

3.3.3 Evaluation

In this section, computer simulation results and comparisons with [42] are presented to demonstrate the performance of the proposed approach. Consider a region D in a 1000 by 1000 m^2, see Figure 3.5. We take $Z_{max} = 120$ m, $Z_{min} = 30$ m[1] and $\alpha = \frac{\pi}{2}$.

---

1 https://bit.ly/2HfWUC4.

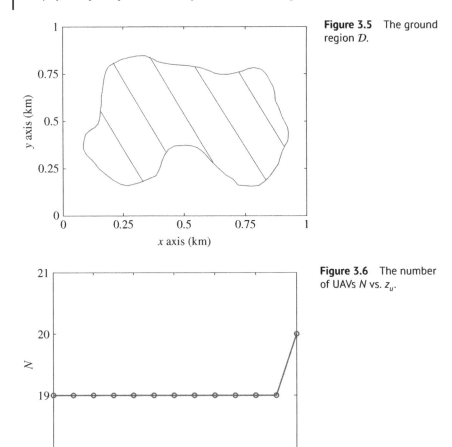

Figure 3.5 The ground region D.

Figure 3.6 The number of UAVs N vs. z_u.

First, the proposed approach is applied to find the minimal number and the positions of UAVs. Starting from $z_u = Z_{max}$ in Step **B1**, 3400 configurations of λ, x_0, y_0 are randomly generated (the resolution for λ takes $\frac{\pi}{30}$ and that for x_0 and y_0 takes 5 m), and then the algorithm **A1–A6** is applied to each of them. The deployment which corresponds to the minimum number of UAVs is recorded. In Step **B2**, ϵ is set as 1 m. With the decrease of z_u, the number of UAVs keeps 19; while when z_u turns to 108 m, the number of UAVs increases to 20, see Figure 3.6. Therefore, to fully cover D, 19 UAVs are required, and their lowest altitude is 109 m. The corresponding deployment of the 19 UAVs at the altitude 109 m is demonstrated in Figure 3.7.

For comparison, the method proposed by Savkin and Huang [42] is applied to the same case. Given a number of UAVs, the review [42] presents a Voronoi cell-based approach to maximize the coverage quality, which is modeled as

Figure 3.7 Deployment of 19 UAVs at 109 m by the proposed approach. The dash circles are the coverage areas of UAVs.

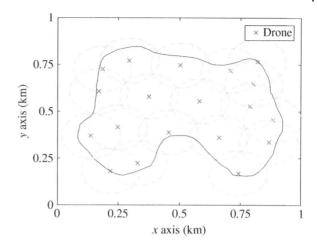

minimizing the weighted distance from the points in the region to UAVs. To make this method work for the current scenario, the points inside $\mathcal{D}$ are given the weight 1, while others are given the weight 0. For the altitude $z_u = 109$ m, the proposed approach requires 19 UAVs to completely cover $\mathcal{D}$. For the compared method, N is first set as 19, and the method [42] is used to compute the corresponding deployment. The UAVs are deployed at the centers of the Voronoi cells. To completely cover a Voronoi cell, the coverage range*2pc of a UAV should be a circle and the radius equals the distance from the farthest point in the cell to the center. Therefore, to cover all the cells at the same altitude, the coverage radius of UAVs should be the maximum one among all the radii of the Voronoi cells. For $N = 19$, 20 sets of simulations with different initializations are conducted, and the smallest radius is 114 m, which corresponds to the altitude 114 m (as $\tan(\frac{\alpha}{2}) = 1$). Comparing this with Figure 3.7, it can be seen that there are more overlaps between circles in Figure 3.8a than in Figure 3.7, which results in the fact that to cover the same area with the same number of UAVs, the method of [42] needs larger radius (i.e. higher altitude) than the proposed approach. Now, the number of UAVs is decreased by one, and the above procedures are repeated. The deployment as shown in Figure 3.8b is obtained where the coverage radius is 108 m (which is comparable to 109 m), so as the altitude of UAVs. Therefore, in covering a given area $\mathcal{D}$ at the same (similar) altitude, the proposed method outperforms [42] in terms of the required number of UAVs.

Now, Proposition 3.2 is illustrated through the following simulations. The area of $\mathcal{D}$ is about 0.41 km^2. By setting the altitude z_u as 109 m, the right hand of Eq. (3.5) is 13.2. The parameter γ takes values from 1 to 10. For each γ, 20 configurations of λ, x_0, and y_0 are randomly generated. For all these configurations, the triangulations are constructed, and the UAVs are deployed following the algorithm **A1–A6**. For each γ, the deployment that has the lowest number of UAVs is recorded. The relationship between $\frac{N(\gamma)}{\gamma^2}$ and γ is shown in Figure 3.9.

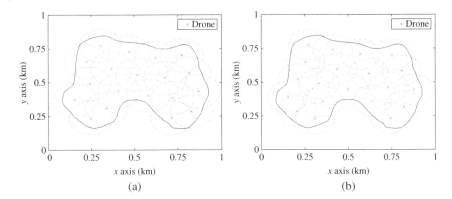

Figure 3.8 The deployments by the algorithm of [42]. (a) $N = 19$ and the UAVs are at the altitude 114 m. (b) $N = 20$, and the UAVs are at the altitude 108 m. Source: Adapted from Savkin and Huang [42].

Notice that with the increase of γ, the region $\mathcal{D}_\gamma$ becomes larger. For a fixed altitude z_u, the coverage radius of the UAVs r is fixed, so as r_t. Thus, the number of vertices in the constructed triangulation increases significantly with γ, which increases the simulation time dramatically. So the simulations are conducted only for γ up to 10. It can be seen from Figure 3.9 that $\frac{N(\gamma)}{\gamma^2}$ converges to 13.2 as it predicted by Proposition 3.2. For comparison, the method of [42] is also applied to the cases with different γ. Again, the deployments which have the smallest numbers of UAVs among 20 simulations for each γ are recorded. The corresponding $\frac{N(\gamma)}{\gamma^2}$ versus γ is shown in Figure 3.9 as well. Clearly, the proposed approach outperforms the deployment algorithm of [42] in terms of number of UAVs.

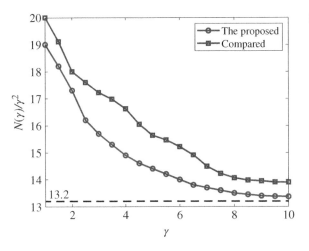

Figure 3.9 $\frac{N(\gamma)}{\gamma^2}$ vs. γ.

3.4 UAV Deployment for Surveillance of Uneven Ground Areas

Beyond Section 3.3, in this section, we consider the scenario where the area of interest is uneven. The mathematical description of the deployment problem is stated in Section 3.4.1, and the deployment algorithm is proposed in Section 3.4.2.

3.4.1 Problem Statement

In addition to the surveillance of a flat area discussed in Section 3.3, this section considers the more challenging situation, i.e. the surveillance of an uneven area.

Here, (x, y) is Cartesian coordinates on the ground plane, and z be the coordinate axis perpendicular to the ground plane, which are the same as in Section 3.3. A terrain is a graph of a function $F(x, y)$ that assigns to every point (x, y) on the ground plane an elevation $z = F(x, y)$. The case $F(x, y) = 0$ for all (x, y) corresponds to a perfectly flat (even) terrain. Let $\hat{D} := \{(x, y, F(x, y))\}$ be the area of the terrain, where $(x, y) \in D$, and D is a given bounded subset of the ground plane $z = 0$ as defined in Section 3.3. The objective is to deploy a number of aerial UAVs to monitor the corresponding area of the terrain, that is the set $\hat{D}$. Each UAV can be deployed only at some points (x_u, y_u, z_u) satisfying (3.1). Furthermore, let ρ_{ij} be the distance between the UAVs i and j, and ρ_i be the minimum distance between the UAV i and the terrain. The following safety constraints should hold:

$$\rho_{ij} \geq c_1, \quad \rho_i \geq c_2, \tag{3.7}$$

where $c_1 > 0$ and $c_2 > 0$ are some given safety margins. The requirements (3.7) allow to avoid collisions of UAVs with the terrain and each other, which is very important in navigation and deployment, see e.g. [39–41, 43].

With a given observation angle $0 < \alpha < \pi$, which defines the visibility cone of each UAV, a UAV with the coordinates (x_u, y_u, z_u) can only see points (x, y, z) of the terrain that are inside of the circle of radius

$$r(z) := \tan\left(\frac{\alpha}{2}\right)(z_u - z) \tag{3.8}$$

centred at (x_u, y_u, z_u) where $z < z_u$. Note that the radius $r(z)$ depends on the altitude z, a bit different from (3.2). A point P on the terrain is visible from a UAV located at the point D if P is inside of the visibility cone of the UAV, and there is not any other point of the terrain on the straight line segment (D, P), see Figure 3.10.

Definition 3.3 Deployment of a number of UAVs is said to be covering if the constraints (3.1), (3.7) are satisfied and every point on the terrain region $\hat{D}$ is visible to at least one of the UAVs.

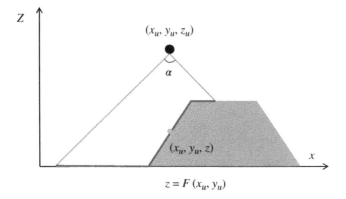

Figure 3.10 The region a UAV can see.

Terrain Model: This review considers the following class of very uneven terrains. First, it is assumed that the set D is a polygon with n vertices. Furthermore, let $D_1, \ldots, D_k$ be some nonoverlapping polygons inside D with $n_1, \ldots, n_k$ vertices, respectively. These polygons represent very uneven areas inside D such as buildings, hills, mountains, walls. These areas are modeled as polytopes as follows: Let $\mathcal{E}_1, \ldots, \mathcal{E}_k$ be some other polygons inside the polygons $D_1, \ldots, D_k$, respectively. It is assumed that each polygon $\mathcal{E}_i$ has the same number n_i of vertices with the polygon D_i. It should be pointed out that $D, D_1, \ldots, D_k$ and $\mathcal{E}_1, \ldots, \mathcal{E}_k$ may be nonconvex. Each D_i represents the "base" face of the polytope modeling the corresponding very uneven area, whereas $\mathcal{E}_i$ represents its "top" face. Furthermore, it is assumed that this polytope has k_i "side" faces. Each "side" face is a convex quadrilateral with one side that is a side of D_i and with the opposite side that is a side of $\mathcal{E}_i$, see Figure 3.11. These quadrilaterals are called side quadrilaterals. Notice that the case when some side of $\mathcal{E}_i$ is a subinterval of D_i means that

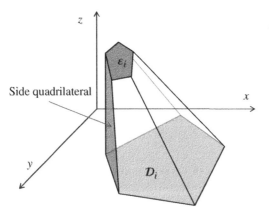

Figure 3.11 An illustration of a very uneven area.

the corresponding side face is a vertical wall. Also, the planes of the "top" faces are not assumed to be parallel to the ground plane $z = 0$, see Figure 3.11. However, they are assumed to be not orthogonal to it. For example, a standard rectangular building is modeled by two identical rectangles $\mathcal{E}_i$ and $\mathcal{D}_i$. Moreover, it is assumed that the rest of polygon $\mathcal{D}$ that is outside of the very uneven areas $\mathcal{D}_1, \dots, \mathcal{D}_k$, is "relatively even." More precisely, suppose that the following assumption holds.

Assumption 3.1 Any point (x, y, z) of the terrain outside the "very uneven" areas satisfies

$$|z| \leq \epsilon, \tag{3.9}$$

where $\epsilon > 0$ is some given constant. Moreover, any two such points outside of the very uneven areas of the terrain satisfy the following constraint: the angle between the straight line connecting these two pints and the ground plane is less than $\beta := \left(\frac{\pi}{2} - \frac{\alpha}{2} \right)$.

It is also assumed that the following technical assumptions usually hold in practice.

Assumption 3.2 The inequality $\epsilon + c_2 \leq Z_{min}$ holds.

Assumption 3.3 The altitude of any point (x, y, z) corresponding to any vertex of any $\mathcal{E}_i$ satisfies $z > \epsilon$.

The problem studied in this chapter that may be called **the UAV version of the 3D AGP** can be stated as follows.

Problem Statement: What is the minimum number of aerial UAVs for which covering deployment exists and where should they be deployed? Moreover, it is preferred not only to deploy the minimum number of UAVs but also deploy the UAVs at as low altitudes as possible to make them closer to the observed region of the terrain.

3.4.2 Deployment Algorithm

The deployment algorithm requires some geometric constructions. Let $\mathcal{P}$ be the polygon that is obtained from the polygon $\mathcal{D}$ with all the interior points of the polygons $\mathcal{D}_1, \dots, \mathcal{D}_k$ taken away. It is a nonconvex polygon with k "holes" and $n + n_1 + \cdots + n_k$ vertices. The proposed algorithm consists of the following steps:

Step B1: Choose k nonintersecting diagonals of the polygon $\mathcal{P}$ that cut it into a polygon Q without "holes." More precisely, each of the k diagonals connects either two vertices of $\mathcal{D}_i$ and $\mathcal{D}_j$ for some $i \neq j$ or two vertices of $\mathcal{D}_i$ and $\mathcal{D}$. Now, consider each of the k diagonals as two different sides of

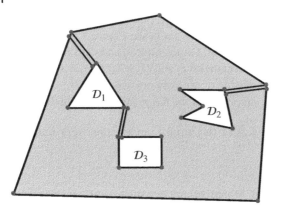

Figure 3.12 The construction of polygon Q by adding k non-intersecting diagonals to the polygon P.

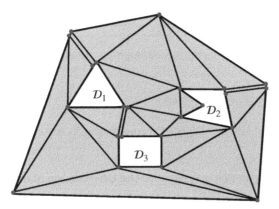

Figure 3.13 The triangulation T of the polygon Q.

the polygon Q, and each vertex of this diagonal as two different vertices of the polygon Q, see Figure 3.12. Hence, the polygon Q is a polygon without "holes" with $n + n_1 + \cdots + n_k + 2k$ vertices.

Step B2: Make some triangulation T of the polygon Q. It means that Q is cut into triangles of vertices of which are vertices of the polygon Q, and all sides are either sides of Q or its nonintersecting diagonals, see Figure 3.13. Since Q has $n + n_1 + \cdots + n_k + 2k$ vertices and no "holes," the number of triangles in any such a triangulation T is $n + n_1 + \cdots + n_k + 2k - 2$.

Step B3: Build a triangulation $\hat{T}$ by enlarging the triangulation T by adding some triangles as follows: For any i, add all the vertices e_i^j of $\mathcal{E}_i$ to T. So $n_1 + n_2 + \cdots + n_k$ vertices have been added to the triangulation. $n_1 + n_2 + \cdots + n_k$ new triangles as added as follows: For any added vertex e_i^j, add the triangle $(d_i^j d_i^{j+1} e_i^j)$, where d_i^j, d_i^{j+1} are the corresponding vertices of D_i, see Figure 3.14.

Figure 3.14 The construction of the triangulation $\hat{\mathcal{T}}$. A, B, C are vertices of $\mathcal{D}_1$ and A′, B′, C′ are vertices of $\mathcal{E}_1$.

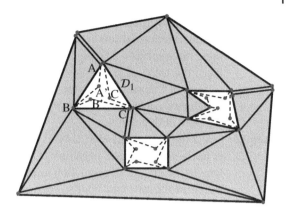

Step B4: Paint the vertices of the triangulation $\hat{\mathcal{T}}$ into three different colors so that any triangle of the triangulation has vertices of three different colors [44]. The following is to prove that such three-coloring exists and give a method to build it. First, build the so-called "dual graph" of the triangulation $\hat{\mathcal{T}}$ in which the vertices correspond to the triangles of the triangulation $\hat{\mathcal{T}}$, and two vertices are connected by an edge if and only if the corresponding triangles have a common side. It is obvious that this dual graph is a tree. Indeed, if it is not a tree, then it has a cycle, hence, there exists a hole inside of this cycle in the polygon Q, which contradicts to the fact that Q has no holes. Furthermore, since this dual graph is a tree, it should have a hanging vertex (a vertex with only one edge). Let us denote this vertex V_1 and take it away from the graph. The remaining graph is a tree again; hence, it has a hanging vertex. Let us denote this vertex V_2 and take it away the graph. This operation is done step by step until last vertex V_M of the dual graph is got. Now, paint all the vertices of the triangulation $\hat{\mathcal{T}}$ into three colors as follows. First, paint three vertices of the triangle corresponding to V_M into three different colors. Then for $i = 1, \ldots, M - 1$, if all the vertices of all the triangles corresponding to $V_M, \ldots, V_{M-i+1}$ have been painted, take the triangle corresponding to V_{M-i}. By the above construction, this triangle has a common side with one of the triangles corresponding to $V_M, \ldots, V_{M-i+2}, V_{M-i+1}$ and no common sides with all other of these triangles. The two vertices corresponding to the common side have already been painted (they are vertices of the triangle corresponding to some V_j, where $M \geq j \geq M - i + 1$). Now paint the third vertex into the third remaining color. So step by step all the vertices of the triangulation $\hat{\mathcal{T}}$ have been painted.

Step B5: The set of vertices of the triangulation $\hat{\mathcal{T}}$ is now divided into three nonintersecting subsets corresponding to three different colors. Among these three subsets, take the one with the minimum number of vertices. Place the UAVs at points with (x, y) coordinates corresponding to the vertices of this subset. If some vertex is a vertex of a cutting diagonal, which now corresponds to two vertices of the polygon Q, only one UAV is placed at that vertex.

Step B6: Select the altitude of each UAV as follows: If the corresponding vertex is not a vertex of any polygons $\mathcal{D}_i, \mathcal{E}_i$, let d_l be the maximum of the lengths of all the triangulation triangles sides for which this vertex is one of the two end points. Then place a UAV at this vertex at the altitude

$$
z_u := \max \left\{ Z_{min}, \epsilon + \frac{d_l}{\tan\left(\frac{\alpha}{2}\right)} \right\}. \tag{3.10}
$$

If the corresponding vertex is a vertex of some polygons $\mathcal{D}_i, \mathcal{E}_i$, let $\hat{d}_l$ be the maximum of the lengths of all the triangulation triangles sides for which this vertex is one of the two end points and the distances to all the vertices of the two side quadrilaterals for which this point is a vertex. Moreover, let a be the maximum altitude of all the terrain points corresponding to the vertices of the two side quadrilaterals for which this point is a vertex. Furthermore, let d_m be the maximum distance from this vertex to the vertices of the corresponding top side polygon $\mathcal{E}_i$, and b be the maximum altitude of all the terrain points corresponding to the vertices of $\mathcal{E}_i$ and the two corresponding side quadrilaterals. Also, let $\hat{d}_m := \max\{\hat{d}_l, d_m\}$. Then for each i, select one vertex that is a vertex of some polygon $\mathcal{D}_i$ or $\mathcal{E}_i$ and place a UAV at this vertex at the altitude

$$
z_u := \max \left\{ Z_{min}, b + c_2 + \frac{\hat{d}_m}{\tan\left(\frac{\alpha}{2}\right)} \right\}. \tag{3.11}
$$

For all other selected vertices, that are vertices of either polygon $\mathcal{D}_i$ or $\mathcal{E}_i$, place a UAV at this vertex at the altitude

$$
z_u := \max \left\{ Z_{min}, a + c_2 + \frac{\hat{d}_l}{\tan\left(\frac{\alpha}{2}\right)} \right\}. \tag{3.12}
$$

Remark 3.9 It should be pointed out that in the proposed algorithm **B1–B6**, this approach always considers triangulations on the ground plane $z = 0$. So all vertices of triangulations are projections of the points on the actual uneven terrain, and the lengths of triangulations sides and quadrilaterals diagonals are also taken on the plane, not on the actual terrain.

Remark 3.10 In the case, when two selected vertices on the triangulation correspond to the same point of the actual terrain, which can happen when either there are two vertices from one which is the end point of a cutting diagonal in **B1** or a vertex of D_i coincides with a vertex of $\mathcal{E}_i$, place just one UAV at the point corresponding these two vertices.

Remark 3.11 The step **B4** of the algorithm is based on the three-coloring method of [12].

For the main result, the following assumptions are needed.

Assumption 3.4 The UAVs altitudes z_u defined by (3.10), (3.11), (3.12) satisfy $z_u \leq z_{max}$.

Assumption 3.5 The length d of any side of any triangle of the triangulation $\mathcal{T}$ satisfies $d \geq c_1$.

The following notation is also needed. For any number $x \geq 0$, $\lfloor x \rfloor$ denotes the integer part of x, i.e. the maximal integer i such that $i \leq x$.

Now, it is the position to state the main result of this section.

Theorem 3.1 A number N of UAVs are deployed by the algorithm **B1–B6**. Suppose that Assumptions 3.1–3.5 hold. This deployment is covering and

$$N \leq \left\lfloor \frac{n + 2n_1 + \cdots + 2n_k + 2k}{3} \right\rfloor. \tag{3.13}$$

Proof. It is obvious that the number of vertices in the constructed triangulation $\hat{\mathcal{T}}$ is $n + 2n_1 + \cdots + 2n_k + 2k$. Since all these vertices are painted into three colors and take a color with the minimum number of vertices, the number of these vertices N satisfies (3.13). Now, prove that this deployment of UAVs is covering. Indeed, by the construction, any triangle of $\hat{\mathcal{T}}$ has a UAV deployed at one of its three vertices. Furthermore, any point outside of the very uneven areas belongs to one of the triangles of the triangulation $\mathcal{T}$, and the UAVs altitudes (3.10)–(3.12) and Assumption 3.1 guarantee that this point is visible from the UAV that is located at one of the three vertices of this triangle. Moreover, it obviously follows from the construction that for any side face of any very uneven area, there is a UAV deployed at one of its four vertices. The UAV altitude selection rules (3.11) and (3.12) guarantee that any point of the side face is visible from a UAV located at one of these four vertices. Furthermore, it is obvious that any point of the top face of any very uneven area is visible from the UAV with altitude selected by (3.12). Moreover, for any UAV location (x_u, y_u, z_u), by the construction $(x_u, y_u) \in D$, it follows from (3.10) to (3.12) that $z_u \geq Z_{min}$, and Assumption 3.4 guarantees that $z_u \leq Z_{max}$. Therefore, the requirements (3.1) hold. Finally, (3.10)–(3.12), and Assumptions 3.2–3.5 imply that the requirements (3.7) are satisfied. This completes the proof of Theorem 3.1. □

3.4.3 Evaluation

This section demonstrates how the proposed approach works through a case study using MATLAB. Consider a 20 m by 20 m square area of interest shown in Figure 3.15a with $k = 3$ very uneven area (a 3D view of them is available in Figure 3.15f). It can be seen from the top view in Figure 3.15a that $n_1 = 3, n_2 = 4$, $n_3 = 5$. The parameters are set as $c_1 = 1$ m, $c_2 = 0.5$ m, $\epsilon = 0.2$ m, $Z_{min} = 4$ m, and $\alpha = \frac{\pi}{2}$. First, $k = 3$ nonintersecting diagonals are added to construct the polygon Q, see Figure 3.15b. It is worth pointing out that there is more than one option to insert this kind of diagonals and only one option is demonstrated here. Clearly, according to the proposed constructed, these diagonals build a Q with no holes. Then, following Step **B2** and **B3**, the triangulation $\hat{\mathcal{T}}$ is constructed as shown in Figure 3.15c. This triangulation has 34 $(n + 2n_1 + 2n_2 + 2n_3 + 2k)$ vertices and 32 $(n + 2n_1 + 2n_2 + 2n_3 + 2k - 2)$ triangles. By selecting a starting triangle, all the triangles are numbered in Figure 3.15c. Then, the dual graph of the triangulation $\hat{\mathcal{T}}$ is constructed as shown in Figure 3.15d. The dual graph has 32 vertices corresponding to the 32 triangles of the triangulation $\hat{\mathcal{T}}$. Following Step **B4**, all the vertices of the dual graph are numbered from 1 to 32 and they are painted in three colors (black, dark gray and light gray). The numbers of vertices with these colors are 13, 10, and 11, respectively, see Figure 3.15e. Hence, the subset of vertices with the color of dark gray is selected. Obviously,

$$N = 10 \leq \left\lfloor \frac{34}{3} \right\rfloor = 11.$$

This also shows that the estimate of Theorem 3.1 is quite close. Finally, the altitudes of the UAVs are computed by (3.10)–(3.12), and the deployment of the aerial UAVs is shown in Figure 3.15f.

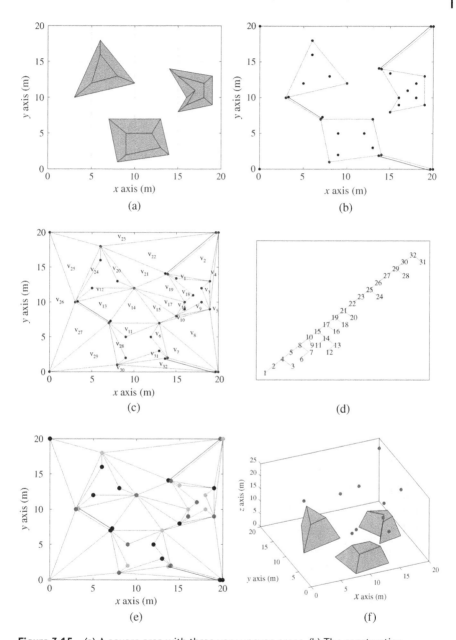

Figure 3.15 (a) A square area with three very uneven areas. (b) The construction of polygon Q. (c) The construction of triangulation $\hat{\mathcal{T}}$. (d) The dual graph of $\hat{\mathcal{T}}$ with the vertices numbered from 1 to 32. (e) Painting the vertices. (f) Deployment of aerial UAVs in 3D space.

3.5   2D UAV Deployment for Ground Target Surveillance

In addition to the area surveillance considered in Sections 3.3 and 3.4, this section and the following one consider the target surveillance problems. Moreover, the targets may move. In Section 3.5.1, we formulate the problem of interest, and in Section 3.5.2, we present the 2D deployment method.

3.5.1   Problem Statement

Consider a monitoring and surveillance system consisting of n UAVs labeled $i = 1, 2, \ldots, n$ and k GNs labeled $j = 1, 2, \ldots, k$. The UAVs are deployed at the same altitude. Again, D is some given bounded closed set with smooth boundary which is called the UAV deployment region. It is assumed that they can be deployed only at some points $P_1, P_2, \ldots, P_n$ of this region, i.e. $P_i \in D$ for all i. D may have "holes" that may correspond, for example, to tall buildings or some other areas over which it is not allowed to deploy UAVs. The fixed locations of GNs are $Q_1, Q_2, \ldots, Q_k$.

To transmit the detected events to a GN without a significantly delay, any UAV is required to link to a GN. Such a link is characterized by a given communication graph $\mathcal{G}$. $\mathcal{G}$ has $n + k$ vertices $v_1^u, v_2^u, \ldots, v_n^u, v_1^g, \ldots, v_k^g$ that correspond to n UAVs and k GNs. Some of the UAVs and GNs should be close enough to each other to be able to communicate according to the communication graph $\mathcal{G}$. For instance, in the illustrative example shown in Figure 3.16, the UAV on the right hand side is connected to the UAV on the left, and the latter is also connected to a GN.

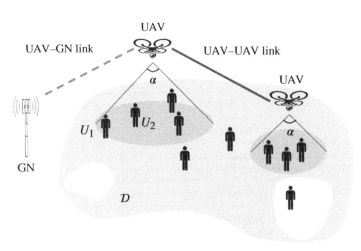

Figure 3.16   Using UAVs for target surveillance.

The deployment of UAVs should satisfy the following constraints:

$$|P_i, P_h| \leq R_1 \tag{3.14}$$

if the UAVs v_i^u, v_h^u are connected by an edge in $\mathcal{G}$; and

$$|P_i, Q_j| \leq R_2 \tag{3.15}$$

if UAV v_i^u and GN v_j^g are connected by an edge in $\mathcal{G}$. Here $R_1 > 0$ and $R_2 > 0$ are some given constants describing the communication range between UAVs and a UAV and a GN, Q_j is the location of GN j, and $|\cdot, \cdot|$ denotes the standard Euclidean distance between two points.

Furthermore, the environment to be monitored is described by a bounded closed region $\mathcal{V}$ and a piecewise continuous function $\rho(p) \geq 0$ defined for all $p \in \mathcal{V}$, that are called the event region and the probability density function. The function $\rho(p)$ describes the probability of an event of interest at a particular point p at a given time. Such a probability density function can be constructed based on statistic data. A typical example of the probability density function is the crowd density function in the problem of crowd monitoring during occasional events such as visiting the temple fairs in Beijing during the Chinese Spring Festival, which can precomputed easily based on previous years statistics.

In surveillance and monitoring applications, aerial UAVs are equipped with some sensors, such as cameras. A common scenario is when each UAV is equipped with a ground facing camera with some visibility angle λ. The camera can see a circular area on the ground with the radius $r := z_u \tan(\frac{\lambda}{2})$ following (3.2), where z_u is the UAV's altitude. Introduce $\mathcal{B}(P_1, \ldots, P_n) \subset \mathcal{V}$ as the set of all points $p \in \mathcal{V}$ such that $|p, P_i| \leq r$ for some i. The QoC by the UAVs placed at points $P_1, \ldots, P_n$ is described by the function:

$$\alpha(P_1, \ldots, P_n) := \int_{p \in \mathcal{B}(P_1, \ldots, P_n)} \rho(p) dp. \tag{3.16}$$

Now, the considered problem can be formulated as follows:

$$\max \int_{p \in \mathcal{B}(P_1, \ldots, P_n)} \rho(p) dp \tag{3.17}$$

subject to

$$|P_i, P_h| \leq R_1, \text{ if } v_i^u \text{ and } v_h^u \text{ are connected,} \tag{3.18}$$

$$|P_i, Q_j| \leq R_2, \text{ if } v_i^u \text{ and } v_j^g \text{ are connected.} \tag{3.19}$$

The objective function $\alpha(P_1, \ldots, P_n)$ defined for all $P_1, \ldots, P_n \in \mathcal{D}$ is a continuous function on a compact set, hence, the maximum in the constrained optimization problem (3.27) is achieved at some $P_1^0, \ldots, P_n^0 \in \mathcal{D}$ [45]. Therefore, there exists a nonempty set of local maximum, and the global maximum is achieved at one of them.

3.5.2 Proposed Solution

To address the considered problem, the following notations will be used. Let $C(P_i, r)$ denote the circle of radius r centered at the point P_i. Moreover, for any point $Y \in C(P_i, r)$, $\hat{Y}$ denotes the opposite point of the circle $C(P_i, r)$, see Figure 3.17a. Furthermore, for any point $Y \in C(P_i, r)$, $S(Y, P_i, r)$ denotes the semicircle of the circle $C(P_i, r)$ such that the point Y is at the middle of it, see Figure 3.17a. It is clear that $S(\hat{Y}, P_i, r)$ is the semicircle of $C(P_i, r)$ that is opposite to $S(Y, P_i, r)$. Also, let $\hat{C}(P_i, r)$ be a subset of the circle $C(P_i, r)$ consisting of all points $p \in C(P_i, r)$ such that $p \in \mathcal{U}$ and $|p, P_h| > r$ for all $h \neq i$. Analogously, $\hat{S}(Y, P_i, r) := S(Y, P_i, r) \cap \hat{C}(P_i, r)$. Correspondingly, $\hat{S}(\hat{Y}, P_i, r) := S(\hat{Y}, P_i, r) \cap \hat{C}(P_i, r)$.

Now, introduce the function $\beta(Y)$ defined for all $Y \in C(P_i, r)$ as follows:

$$\beta(Y) := \int_{p \in \hat{S}(Y, P_i, r)} \rho(p)dp - \int_{p \in \hat{S}(\hat{Y}, P_i, r)} \rho(p)dp. \tag{3.20}$$

$\beta(Y)$ gives the benefit if the UAV moves toward Y a bit, because when the UAV moves, it will newly cover the semicircle with Y in the middle while lose the other semicircle.

Definition 3.4 A circle $C(P_i, r)$ is said to be perfect if $\rho(Y) = \rho(\hat{Y})$ for all opposite points $Y, \hat{Y}$ such that $Y, \hat{Y}$ both belong to $\hat{C}(P_i, r)$, and $\rho(Y) = 0$ for all $Y \in \hat{C}(P_i, r)$ such that $\rho(\hat{Y})$ does not belong to $\hat{C}(P_i, r)$.

Proposition 3.3 If $(P_1^0, P_2^0, \dots, P_n^0)$ is a local maximum in the constrained optimization problem (3.27), (3.18), (3.19), then for any $i = 1, \dots, n$, at least one of the following four conditions holds:

C1: $|P_i^0, P_h^0| = r_1$ for some h such that the corresponding UAVs v_i^r, v_h^r are connected by an edge in $\mathcal{G}$;

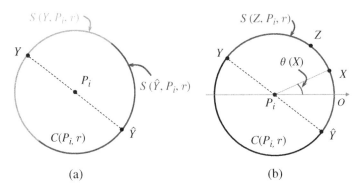

Figure 3.17 Construction of $S(Y, P_i, r)$, $S(\hat{Y}, P_i, r)$, and $\theta(X)$.

C2: $|P_i^0, Q_j| = r_2$ *for some j such that the corresponding UAV v_i^r and the corresponding GN v_j^b are connected by an edge in $\mathcal{G}$;*

C3: P_i^0 *belongs to the boundary of the UAV deployment region $\mathcal{D}$;*

C4: *The circle $C(P_i^0, r)$ is perfect.*

Proof. We prove that if $(P_1^0, P_2^0, \ldots, P_n^0)$ is a point of local maximum and for some i conditions **C1–C3** do not hold, then condition **C4** holds. First, we prove that $\beta(Y) = 0$ for any $Y \in C(P_i^0, r)$, where $\beta(Y)$ is defined by (3.20). Indeed, it is obvious that $\beta(Y) = -\beta(\hat{Y})$, hence, if $\beta(Y) = 0$ does not hold for some point, then $\beta(Y) > 0$ for some Y. Consider now the straight line defined by the points P_i^0 and Y and introduce the coordinate x along this straight increasing from P_i^0 to Y so that $x = 0$ corresponds to P_i^0. Then $\frac{d\alpha(x)}{dx} = \beta(Y) > 0$ for $x = 0$. Therefore, if for some small movement of the point P_i^0 toward Y, the value $\alpha(P_i^0)$ will increase, and since **C1–C3** do not hold, the constraints (3.18) and (3.19) will still hold, and P_i^0 will still be in $\mathcal{D}$. This means that if all four conditions **C1–C4** do not hold, we can make a small motion of P_i^0 so that the function (3.27) increases. Hence, $(P_1^0, P_2^0, \ldots, P_n^0)$ is not a local maximum. Now, for opposite points $Y, \hat{Y}$, consider a semicircle $S(Z, P_i, r)$ such that $Y, \hat{Y}$ are its boundary points. Introduce some directed line (P_i^0, O). For any point X of the circle $C(P_i, r)$, let $\theta(X)$ denote the angle between the lines (P_i^0, O) and (P_i^0, X) measured in the counterclockwise direction, see Figure 3.17b. Furthermore, let $\hat{\beta}(\theta(Z)) := \beta(Z)$ for any point Z. As we have already proved, $\hat{\beta}(\theta(Z)) \equiv 0$, hence $\frac{d\hat{\beta}(\theta)}{d\theta} \equiv 0$. On the other hand, it is obvious that for $\theta = \theta(Z)$, we have $\frac{d\hat{\beta}(\theta)}{d\theta} = 2(\rho(Y) - \rho(\hat{Y}))$ if $Y, \hat{Y}$ both belong to $\hat{C}(P_i, r)$ and $\frac{d\hat{\beta}(\theta)}{d\theta} = 2\rho(Y)$ if $Y \in \hat{C}(P_i, r)$ and $\hat{Y}$ does not belong to $\hat{C}(P_i, r)$. The statement of Proposition 3.3 immediately follows from this. $\qquad\square$

Let $\mathcal{P}_j$ and $\mathcal{Q}_j$ denote the set of points P_i satisfying the constraint (3.18) and (3.19) for a particular j. Obviously, such sets are balls and their boundaries are circles of radius r_1 and r_2, respectively. From some initial position $P_1, P_2, \ldots, P_n \in \mathcal{D}$ satisfying the constraints (3.18) and (3.19), each UAV operates according to Algorithm 3.1.

Let $N = sn + i$, where $s = 0, 1, 2, \ldots$ and $i = 1, 2, \ldots, n$. At each step N, the point P_i continuously moves as follows:

1. if P_i is inside $\mathcal{D} \cap \mathcal{P}_j \cap \mathcal{Q}_j$, i.e. Line 3, find a point Y^0 on $C(P_i, r)$ that maximizes $\beta(Y)$. Obviously, $\beta(Y^0) \geq 0$. P_i continuously moves along a trajectory we build so that the vector $P_i Y^0$ is tangent to this trajectory. P_i stops when it reaches a point beyond which it is impossible to continue the trajectory of P_i so that P_i remains in $\mathcal{D}$, and the conditions (3.18), (3.19), $\beta(Y^0) \geq 0$ hold;
2. if P_i is on the boundary of one of $\mathcal{D}, \mathcal{P}_j, \mathcal{Q}_j$, i.e. Line 5, then we build a trajectory as in (1) with the circle $C(P_i, r)$ replaced by the semicircle $S(Y, P_i, r)$ cut by the

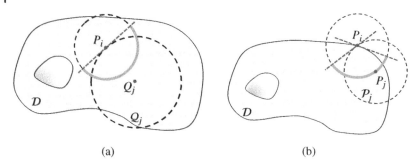

(a)   (b)

Figure 3.18 Candidate sectors (solid curve on the circle centred at P_i) of the circle $C(P_i, r)$ on which we look for a point Y^0 to maximize $\beta(Y)$.

tangent to the boundary of the corresponding set (this guarantees that P_i will not leave $\mathcal{D}, \mathcal{P}_j, \mathcal{Q}_j$, see Figure 3.18a). Let Θ_1 denote such a semicircle;

3. if P_i is on the boundary of two or more of $\mathcal{D}, \mathcal{P}_j, \mathcal{Q}_j$, i.e. Line 7, then we build a trajectory as in (1) with the circle $C(P_i, r)$ replaced by the sector of this circle that is the intersection of the straight angles made by the tangents to the boundaries of the corresponding sets, see Figure 3.18b. Let Θ_2 denote such circle sector.

Algorithm 3.1 The algorithm running at UAV i.

1: **while** $P_i \in \mathcal{D}$, (3.18), (3.19), $\beta(Y_0) \geq 0$ hold **do**
2: Compute $\mathcal{P}_j, \mathcal{Q}_j$.
3: **if** P_i is inside $\mathcal{D} \cap \mathcal{P}_j \cap \mathcal{Q}_j$ **then**
4:   Find $Y_0 \in [0, 2\pi)$ that maximizes (3.20).
5: **else if** P_i is on the boundary of one of the sets $\mathcal{D}, \mathcal{P}_j, \mathcal{Q}_j$ **then**
6:   Find $Y_0 \in \Theta_1$ that maximizes (3.20).
7: **else if** P_i is on the boundary of two or more of the sets $\mathcal{D}, \mathcal{P}_h, \mathcal{Q}_j$ **then**
8:   Find $Y_0 \in \Theta_2$ that maximizes (3.20).
9: **end if**
10: P_i moves towards Y_0 for a unit distance.
11: **end while**

Proposition 3.4 *The proposed algorithm reaches a local maximum of the constrained optimization problem (3.17)–(3.19) in a finite number of steps.*

Proof. The proposed algorithm generates a sequence $(P_1, P_2, \ldots, P_n)_N$. Since this sequence is in some compact set $\mathcal{D}$, it has a subsequence that converges to some $(P_1^0, P_2^0, \ldots, P_n^0)$ [45]. In the movement of P_i, the condition $\beta(Y^0) \geq 0$ always holds. Since the derivative of the function (3.16) along the trajectory we build is equal to $\beta(Y^0)$, the function (3.16) is non-decreasing on the sequence $(P_1, P_2, \ldots, P_n)_N$.

Hence, $(P_1^0, P_2^0, \ldots, P_n^0)$ is a local maximum. Furthermore, it follows from the proposed algorithm that when $(P_1, P_2, \ldots, P_n)$ in some small neighborhood of $(P_1^0, P_2^0, \ldots, P_n^0)$, it will reach $(P_1^0, P_2^0 \ldots, P_n^0)$ and stays there. $\qquad\square$

Now, we analyze the complexity of the algorithm. For one UAV to make movement decision, the algorithm checks conditions (3.18), (3.19), and finds Y^0 according to (3.20). In the worst case, the computational complexities of these operations are $O(n)$, $O(k)$, and $O(n)$, correspondingly. Hence, the computational complexity of making decisions by n UAVs is $O(n^2 + nk)$. To implement the system in a real network, we may need to update positions of UAVs in real time. The UAVs move in a cycle. The UAV i moves according to Algorithm 1, and once it cannot move anymore, it broadcasts a message with its index and position. After receiving the broadcast, the UAV $i + 1$ updates the position record and starts to move; while other UAVs only update the position records but do not move. When the UAV $i + 1$ cannot move anymore, it broadcasts a message and this process repeats. The UAV 1 reacts to the broadcast from the UAV n. Therefore, to make the UAVs collaborate, a broadcast message to let the next UAV know that it is ready to move and updated position records are associated overheads. Thus, the overall overhead is $O(n)$.

3.5.3 Evaluation

Simulation results based on the real dataset Momo are presented. Momo is a social discovery App, and when a Momo user has an update, the information of his ID, timestamp, latitude, and longitude is sent to the server. The Momo dataset contains approximately 150 million such updates in 38 days, from 21 May 2012 to 27 June 2012. The updates by the users in an area in Beijing, China, is considered. Further, a plane with the grid size of 10 m is constructed, and the average number of users falling into each cell over 38 days is counted; see Figure 3.19a. The UAV deployment region D is shown in Figure 3.19a (black lines). The region where UAVs can be deployed are marked.

Figure 3.19b shows how 25 UAVs move to the local optimal positions (solid curves). The communication links are marked by dash lines, and they link only the final positions of UAVs. All UAVs remain within D. For instance, P_1 moves along the boundary of a hole without entering and P_2 moves along the boundary of D without leaving. Further, the coverage ratio ($\alpha(P_1, \ldots, P_n)$/the total number of users in $\mathcal{U}$) during the movement is shown in Figure 3.19c. The coverage ratio converges to the maximum in 179 steps and it is increased by about 20%. For the considered deployment region D, 100% coverage cannot be achieved. For instance, the bottom-right part of the event region $\mathcal{U}$ is out of range of any UAV in D.

To assess the performance, we compare the proposed algorithm with a baseline method [42]. The idea of [42] is to move each UAV toward the center of mass

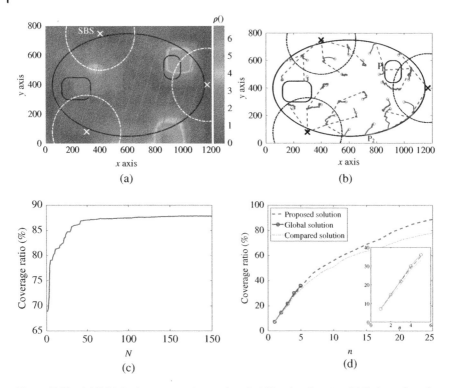

Figure 3.19 (a) UAV deployment region and probability density $\rho(p)$. (b) Trajectories of 25 UAVs. (c) Coverage ratio vs. N. (f) Coverage ratio vs. n.

of the targets it serves. Since it is originally to support wireless communication, we remove the communication model part. The number of UAVs is set from 1 to 25. For each number of UAVs, 20 simulations with different initial positions are conducted and the best results among them are presented in Figure 3.19d. It can be seen that with the increase of n, the gain achieved by the proposed approach against the compared one becomes larger and when $n = 25$, the gap is about 10%. Moreover, it is interesting to study how far the locally optimal solutions obtained by the proposed method are from globally optimal solutions. We adopt the brute-force search (exhaustive search) to find the global solutions for the cases with 1–5 UAVs. For larger number of UAVs, the brute-force search cannot be implemented in a reasonable time, since the computing demand of such exhaustive search algorithm increases exponentially with the number of UAVs, therefore, finding the global optimum by it is possible only for small numbers of UAVs. As shown in Figure 3.19d, for $n = 1, 2, 3$ the proposed solutions are the same with the global optimum; while for $n = 4, 5$, the results obtained by the proposed algorithm are about 1% and 1.5% lower than those optimal solutions.

The proposed algorithm outperforms the compared method and as shown in Figure 3.19d the gain becomes more significant with the increase of n.

3.6 3D UAV Deployment for Ground Target Surveillance

Besides the 2D deployment problem, this section pays attention to the 3D deployment problem. Compared to the 2D deployment problem considered in Section 3.5, the 3D version is more difficult due to the additionally dimension. We state the mathematical problem in Section 3.6.1 and then present the deployment method in Section 3.6.2.

3.6.1 Problem Statement

Consider a system consisting of n UAVs labeled $i = 1, 2, \ldots, n$ and k GNs labeled $j = 1, 2, \ldots, k$. Some UAVs and GNs should be able to communicate to each other according to a given connected graph $\mathcal{G}$ which is called the communication graph. These settings are the same as Section 3.5. Let $d_{safe} > 0$ be a given constant describing the minimum safety distance between UAVs. It is also assumed that $d_{safe} < R_1$. The safety requirement is that at any time the following constraints should hold

$$|P_i, P_h| \geq d_{safe} \tag{3.21}$$

for all $i \neq h$. It is always assumed that at the initial time, the safety condition (3.21) holds for all $i \neq h$. The requirement (3.21) allows to avoid collisions of UAVs with each other, which is very important in navigation and deployment, see e.g. [46–50]. In practice, the constant d_{safe} may be different for different types of UAVs. The value of this constant is selected based on properties of a specific UAV type and should be sufficient to avoid any possible collisions between UAVs, see e.g. [39–41, 43].

There are N ground point-wise targets labeled $1, 2, \ldots, N$ with the coordinates $U_1, U_2, \ldots, U_N$, where $U_j \in \mathbf{R}^2$. Notice that the targets may be outside of the region $\mathcal{D}$. With a given observation angle $0 < \alpha < \pi$, the UAV i with the coordinates (x_i, y_i, z_i) can only see ground targets that are inside of the circle of radius $r_i := \tan(\frac{\alpha}{2})z_i$, centred at (x_i, y_i). Assume that if the distance between the target j and the projection (x_i, y_i) of the UAV i on the ground plane (x, y) is no greater than r_i, then the UAV i knows the coordinates U_j of the target j.

Let $w(d)$ be a given differentiable weighting function such that $w(d)$ and its derivative $w'(d)$ are defined for $d \geq 0$ and satisfy the following conditions:

$$w(d) > 0, \ \forall d \in [0, 1), \quad w(d) = 0, \ \forall d \geq 1;$$
$$w'(0) = w'(1) = 0, \quad w'(d) < 0, \ \forall d \in (0, 1). \tag{3.22}$$

Furthermore, let $g(z)$ be a given differentiable weighting function such that $g(z)$ and its derivative $g'(z)$ are defined for all altitudes $z \in [Z_{min}, Z_{max}]$ and satisfy the following conditions:

$$g(z) > 0, \quad g'(z) < 0, \ \forall z \in [Z_{min}, Z_{max}]. \tag{3.23}$$

Examples of $w(d)$ and $g(z)$ satisfying (3.22) and (3.23) are the following functions:

$$w(d) = \cos(\pi d) + 1, \ \forall d \in [0, 1); \quad w(d) = 0, \ \forall d \geq 1,$$
$$g(z) = \frac{1}{z}, \ \forall z \in [Z_{min}, Z_{max}]. \tag{3.24}$$

Definition 3.5   QoC of target j from UAV i: For target j ($j = 1, 2, \ldots, N$), the QoC from the UAV i is defined by the function $g(z_i)w(\frac{d_i^j}{r_i})$, and d_i^j is the distance between the target j and the projection (x_i, y_i) of the UAV i to the ground plane.

It is clear that $g(z_i)w(\frac{d_i^j}{r_i}) = 0$ if the target is outside of the visibility cone of the UAV i. Moreover, $g(z_i)w(\frac{d_i^j}{r_i})$ is increasing as d_i^j is decreasing; and $g(z_i)w(\frac{d_i^j}{r_i})$ is increasing as z_i is increasing. Hence, the closer the UAV i is to the target j (the smaller the altitude of the UAV is), the bigger $g(z_i)w(\frac{d_i^j}{r_i})$ is. On the other hand, if the altitude of the UAV is fixed, the closer the target is to the boundary of the visibility cone of the UAV, the smaller $g(z_i)w(\frac{d_i^j}{r_i})$ is.

Definition 3.6   QoC of target j: Since a target may be monitored by more than one UAVs at the same time, the QoC of this target is defined as the largest QoC among all the UAVs. For target j, introduce the pair (z^j, d^j) such as

$$g(z^j)w\left(\frac{d^j}{r^j}\right) = \max_{i=1,\ldots,n} g(z_i)w\left(\frac{d_i^j}{r_i}\right), \tag{3.25}$$

i.e. (z^j, d^j) corresponds to the UAV which monitor the target j with the best QoC in sense of the measure $g(z_i)w(\frac{d_i^j}{r_i})$.

Definition 3.7   QoC by the UAVs: For the targets' locations $U_1, U_2, \ldots, U_N$, the QoC by the UAVs placed at points $P_1, P_2, \ldots, P_n$ is described by the following function:

$$W(P_1, \ldots, P_n) := \sum_{j=1,\ldots,N} g(z^j)w\left(\frac{d^j}{r^j}\right). \tag{3.26}$$

It is also assumed that each UAV knows the coordinates of some other UAVs that are close enough to it. For this purpose, it is assumed that each UAV is equipped with a GPS and it knows its own current position. Every a certain period of time, it broadcasts its current position. A UAV can receive at most $n - 1$ messages from other UAVs. Furthermore, each UAV knows for each target that is inside its vision cone via some image processing techniques. With the positions of neighbor UAVs, whether the minimum in (3.25) is achieved at this UAV or not can be known by a UAV.

Now, consider the following optimization problem:

$$W(P_1, \ldots, P_n) \to \max, \tag{3.27}$$

where the maximum is taken for fixed targets' locations $U_1, U_2, \ldots, U_N$ over all $P_1, \ldots, P_n \in D \times [Z_{min}, Z_{max}]$ satisfying the constraints (3.18), (3.19), (3.21).

Problem Statement: The optimization problem under consideration is as follows: for the given targets' locations $U_1, U_2, \ldots, U_N$, the communication graph $\mathcal{G}$, the UAV deployment region $D \times [Z_{min}, Z_{max}]$, the GNs' locations $Q_1, \ldots, Q_k$, the weighting functions $w(d), g(z)$, and the constants r_1, r_2, d_{safe}, move UAVs to locations $P_1, \ldots, P_n \in D$ that maximize the function (3.27) so that the constraints (3.18), (3.19), (3.21) are always satisfied.

3.6.2 Proposed Solution

The following notations are needed to present the proposed solution. The set of N targets is divided into n nonintersecting subsets $\mathcal{V}_1, \mathcal{V}_2, \ldots, \mathcal{V}_n$, where for any $i = 1, 2, \ldots, n$, $\mathcal{V}_i$ is the set of targets j for which the distance between the target j and the projection (x_i, y_i) of the UAV i on the ground plane (x, y) is less than r_i, and the maximum in (3.25) is achieved at this UAV i. In cases, when the maximum in (3.25) is achieved at more than one UAV, the target j belongs to the corresponding set $\mathcal{V}_i$ with the smallest index i among all such UAVs.

Furthermore, for any i, consider an axes in the ground plane (x_i, y_i) departing from the current location of the UAV i into some given direction. Let θ be the angle measured in the counterclockwise direction from this axes. θ takes values in the interval $[0, 2\pi)$. Now, for all $j \in \mathcal{V}_i$, θ_j denotes the angle between the axes and the direction to the location of the target U_j, see Figure 3.20. Let $\beta(\theta)$ be the function defined as

$$\beta(\theta) := -\sum_{j \in \mathcal{V}_i} w'\left(\frac{d^j}{r}\right) \cos(\theta_j - \theta). \tag{3.28}$$

Moreover, let $\gamma(x, y, z)$ be the function defined as

$$\gamma(x, y, z) := g'(z) \sum_{j \in \mathcal{V}_i} w\left(\frac{d^j}{\tan(\frac{\alpha}{2})z}\right) - \frac{g(z)}{\tan(\frac{\alpha}{2})z^2} \sum_{j \in \mathcal{V}_i} d^j w'\left(\frac{d^j}{\tan(\frac{\alpha}{2})z}\right). \tag{3.29}$$

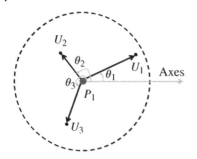

Figure 3.20 Construction of axes and θ_j.

Remark 3.12 The functions (3.28) and (3.29) are defined to navigate UAVs in horizontal direction and vertical direction, respectively. Specifically, $\beta(\theta)$ is proportional to the derivative of the objective function $W(P_1, \ldots, P_n)$ with respective to the distance between the UAV and its covered targets in the horizontal plane; and $\gamma(x, y, z)$ is the derivative of $W(P_1, \ldots, P_n)$ with respective to the z coordinate in the vertical direction. We will show how to use them to navigate UAVs in the proposed algorithm **(D1)–(D3)**.

The function $W(P_1, \ldots, P_n)$ defined for all $P_1, \ldots, P_n \in D \times [Z_{min}, Z_{max}]$ is a continuous function on a compact set, hence, the maximum in the constrained optimization problem (3.27), (3.18), (3.19), (3.21) is achieved at some $P_1^0, \ldots, P_n^0 \in D \times [Z_{min}, Z_{max}]$ [45]. Therefore, there exists a nonempty set of points of local maximum, and the global maximum is achieved at one of them.

Before presenting the proposed algorithm, some properties of the local optimal solution to the problem (3.27), (3.18), (3.19), (3.21) are discussed.

Definition 3.8 Let $(P_1^0, P_2^0, \ldots, P_n^0) \in D \times [Z_{min}, Z_{max}]$ and (x_i, y_i, z_i) are the coordinates of P_i^0 for any i. The point $(P_1^0, P_2^0, \ldots, P_n^0)$ is called a stationary point of the constrained optimization problem (3.27), (3.18), (3.19), (3.21), if for any $i = 1, \ldots, n$, at least one of the following five conditions holds:

S1: $|P_i^0, P_h^0| = R_1$ for some i such that the corresponding UAVs v_i^r, v_h^r are connected by an edge in $\mathcal{G}$;

S2: $|P_i^0, Q_h| = R_2$ for some i such that the corresponding UAV v_i^r and the corresponding GN v_h^b are connected by an edge in $\mathcal{G}$;

S3: $|P_i, P_h| = d_{safe}$ for some $i \neq h$;

S4: The projection (x_i, y_i) of P_i^0 to the ground plane (x, y) belongs to the boundary of D;

S5: $\beta(\theta) = 0$ for all $\theta \in [0, 2\pi)$ where $\beta(\theta)$ is defined by (3.28), and at least one of the following three conditions holds:

    S5A: $z_i = Z_{min}$ and $\gamma(x_i, y_i, z_i) < 0$ where $\gamma(x_i, y_i, z_i)$ is defined by (3.29);

    S5B: $z_i = Z_{max}$ and $\gamma(x_i, y_i, z_i) > 0$;

    S5C: $\gamma(x_i, y_i, z_i) = 0$.

Proposition 3.5 *If $(P_1^0, P_2^0, \ldots, P_n^0) \in \mathcal{D} \times [Z_{min}, Z_{max}]$ is a local maximum in the constrained optimization problem (3.27), (3.18), (3.19), (3.21), then $(P_1^0, P_2^0, \ldots, P_n^0)$ is a stationary point.*

Proof. We prove that if $(P_1^0, P_2^0, \ldots, P_n^0)$ is a point of local maximum and for some i all the conditions **S1–S4** do not hold, then the condition **S5** holds. First, we notice that if $\beta(\theta) < 0$ for some $\theta \in [0, 2\pi)$, then $\beta(\hat{\theta}) > 0$, where $\hat{\theta}$ corresponds to the direction opposite to θ. Therefore, if **S5** does not hold, then $\beta(\theta) > 0$ for some $\theta \in [0, 2\pi)$. Consider now the straight line departing from the point P_i^0 in the direction θ in the plane (x, y) parallel to the ground plane. Moreover, consider the movement of along this line with some constant speed $v > 0$. Then the derivative $\dot{W}(P_1, \ldots, P_n)$ of the function (3.26) along this trajectory satisfies $\dot{W}(P_1, \ldots, P_n) = v\beta(\theta) > 0$. Hence, for some small movement of the point P_i^0 along this straight line, the value $W(P_1, \ldots, P_n)$ will increase, and since **S1–S4** do not hold, the constraints (3.18), (3.19) and (3.21) will still hold, and the coordinates (x_i, y_i) of P_i^0 will still be in $\mathcal{D}$. This means that if the conditions **S1–S4** together with the first of conditions **S5** do not hold, we can make a small motion of P_i^0 so that the function (3.27) increases. Hence, $(P_1^0, P_2^0, \ldots, P_n^0)$ is not a local maximum. Now, we prove that if all condition **S1–S4** do not hold, then at least one of three conditions **S5A–S5C**. Indeed, if all these three conditions and **S1–S4** do not hold, we obviously can make a small motion of P_i along the vertical axis z_i so that the function (3.27) increases and all the constraints of the constrained optimization problem still hold. Hence, $(P_1^0, P_2^0, \ldots, P_n^0)$ is not a local maximum. The statement of Proposition 3.5 immediately follows from this. $\qquad\square$

To describe our algorithm, some more notations are needed. Let $\mathcal{P}_h$ and $\mathcal{Q}_h$ denote the set of points P_i satisfying the constraint (3.18) and (3.19) for a particular h, where i and h are connected in the communication graph $\mathcal{G}$. It is obvious that such sets are balls, and their boundaries are spheres of radius R_1 and R_2, respectively. Moreover, let $\hat{\mathcal{P}}_h$ denote the set of points P_i satisfying the constraint (3.21) for a particular h. It is obvious that such sets are exteriors of spheres of radius d_{safe}. The proposed algorithm is as follows:

D1: Start with some initial points $P_1, P_2, \ldots, P_n \in \mathcal{D} \times [Z_{min}, Z_{max}]$ satisfying the constraints (3.18), (3.19) and (3.21). For any i, continuously move the point P_i as follows:

D2: For each current location of $P_i = (x_i, y_i, z_i)$, move P_i in the plane (x, y) keeping the constant altitude z_i as follows:

    (i) if P_i is an interior point of all the sets $\mathcal{D} \times z_i$, $\mathcal{P}_h$, $\mathcal{Q}_h$, $\hat{\mathcal{P}}_h$, find a value of $\theta_0 \in [0, 2\pi)$ at which the maximum of the function $\beta(\theta)$ defined by (3.28) is achieved. It is obvious that $\beta(\theta_0) \geq 0$. Continuously move the point P_i along a trajectory we build so that the slope of the tangent to this trajectory is always equal to θ_0. Stop this when P_i reaches a point beyond which it is impossible to continue the trajectory of P_i so that (x_i, y_i) remains in D, and the conditions (3.18), (3.19), (3.21), $\beta(\theta_0) \geq 0$ hold;

(ii) if P_i is on the boundary of one of the sets $D \times z_i, P_h, Q_h, \hat{P}_h$, then build a trajectory as in (i) with the circle of possible directions $\theta \in [0, 2\pi)$ replaced by the semicircle cut by the tangent to the boundary of the corresponding set (this guarantees that we will move P_i only in a direction that does not leave $D, P_h, Q_h, \hat{P}_h$, see Figure 3.21b). Such a feasible semicircle is denoted by Θ_1;

(iii) if P_i is on the boundary of two or more of the sets $D \times z_i, P_h, Q_h, \hat{P}_h$, then build a trajectory as in (i) with the circle of possible directions $\theta \in [0, 2\pi)$ replaced by the sector of this circle that is the intersection of the straight

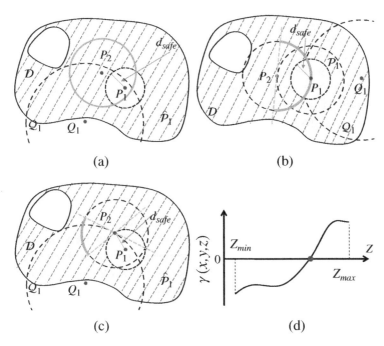

(a)   (b)

(c)   (d)

Figure 3.21 Several typical situations a UAV may meet during the movement. (a) When it is at P_2, it is within R_2 to Q_1 and outside d_{safe} to P_1. The shaded area is the set $\hat{P}_1$, i.e. removing the circular area centered at P_1 with the radius d_{safe} from D. Thus, the coverage circle is the candidate set to find the best angle θ_0. (b) When it is on the boundary of P_1, the candidate set is the semicircle $\Theta_1 = [0, 0.45\pi] \cup [1.45\pi, 2\pi)$. (c) The UAV is on the boundaries of Q_1 and $\hat{P}_1$. Then, the candidate for selecting a future moving direction is the solid sector on the circle centred at P_2: $\Theta_2 = [0.83\pi, 1.2\pi]$. (d) When the UAV moves in the vertical direction, it keeps check on the sign of $\gamma(x, y, z)$, and it always moves toward the direction such that $\gamma(x, y, z)$ becomes closer to 0.

angles made by the tangents to the boundaries of the corresponding sets, see Figure 3.21c. Such a feasible sector is denoted by Θ_2.

D3: If P_i cannot move further in the plane (x, y), we move it along the vertical axis z_i keeping (x_i, y_i) constant as follows: if $\gamma(x_i, y_i, z_i) < 0$, we move P_i down, if $\gamma(x_i, y_i, z_i) > 0$, we move P_i up. So we move P_i until either $\gamma(x_i, y_i, z_i) = 0$ or it is impossible to continue the trajectory so that the conditions (3.18), (3.19), (3.21), $z_i \in [Z_{min}, Z_{max}]$ hold. Then go to **D2**.

To facilitate the understanding of this algorithm, the pseudocode running at each UAV is shown in Algorithm 3.2. Also, corresponding to the cases of **D2** (i), **D2** (ii), **D2** (iii), and **D3**, several typical situations a UAV may meet during the movement are shown in Figure 3.21.

Algorithm 3.2 The algorithm running at UAV i.

1: **while** (3.14), (3.15), (3.21), $\beta(\theta_0) \geq 0$ hold **do**
2:     Compute $\mathcal{P}_h, \mathcal{Q}_h$, and $\hat{\mathcal{P}}_h$.
3:     **if** P_i is inside $D \times z_i \cap \mathcal{P}_h \cap \mathcal{Q}_h \cap \hat{\mathcal{P}}_h$ **then**
4:         Find $\theta_0 \in [0, 2\pi)$ that maximizes (3.28).
5:     **else if** P_i is on the boundary of one of the sets $D \times z_i, \mathcal{P}_h, \mathcal{Q}_h, \hat{\mathcal{P}}_h$ **then**
6:         Find $\theta_0 \in \Theta_1$ that maximizes (3.28).
7:     **else if** P_i is on the boundary of two or more of the sets $D \times z_i, \mathcal{P}_h, \mathcal{Q}_h, \hat{\mathcal{P}}_h$ **then**
8:         Find $\theta_0 \in \Theta_2$ that maximizes (3.28).
9:     **end if**
10:     P_i moves in the horizontal plane along θ_0 for a unit distance.
11: **end while**
12: **while** $\gamma(x_i, y_i, z_i) \neq 0$ and (3.18), (3.19), (3.21), $z_i \in [Z_{min}, Z_{max}]$ hold **do**
13:     Compute $\mathcal{P}_h, \mathcal{Q}_h$, and $\hat{\mathcal{P}}_h$.
14:     **if** $\gamma(x_i, y_i, z_i) < 0$ **then**
15:         P_i moves up for a unit distance.
16:     **else if** $\gamma(x_i, y_i, z_i) > 0$ **then**
17:         P_i moves down for a unit distance.
18:     **end if**
19: **end while**

Proposition 3.6 *The algorithm* **(D1)–(D3)** *converges to a stationary point of the constrained optimization problem (3.27), (3.18), (3.19), (3.21) and the constraints (3.18), (3.19), (3.21) hold at any point of the trajectory we build in* **(D1)–(D3)**.

Proof. The algorithm **(D1)–(D3)** generates a trajectory $(P_1, P_2, \ldots, P_n)$. Since this trajectory is in some compact set $D \times [Z_{min}, Z_{max}]$, it has a subsequence that converges to some $(P_1^0, P_2^0, \ldots, P_n^0)$ [45]. Furthermore, it follows from **(A1)–(A3)** that when $(P_1, P_2, \ldots, P_n)$ in some small neighborhood of $(P_1^0, P_2^0, \ldots, P_n^0)$, it will reach $(P_1^0, P_2^0, \ldots, P_n^0)$ and stays there. Moreover, it is obvious from **(D1)–(D3)**, that it the trajectory cannot be continued beyond a certain point, then this point is stationary. When we move P_i, the condition $\beta(\theta_0) \geq 0$ always holds. Therefore, the function (3.26) is nondecreasing on the trajectory $(P_1, P_2, \ldots, P_n)$. □

More details about the solution to the problem of maximizing (3.28) are presented below. According to (3.24), the term $w'(d_j^{min})$ can be expressed as follows:

$$w'(d^j) = -\frac{\pi}{r} \sin\left(\frac{\pi}{r} d^j\right), \quad \forall d^j \in [0, r). \tag{3.30}$$

Substitute (3.30) into (3.28), one obtains:

$$\begin{aligned}
\beta(\theta) &= \frac{\pi}{r} \sum_{j \in \mathcal{V}_i} \sin\left(\frac{\pi}{r} d^j\right) \cos(\theta_j - \theta) \\
&= \frac{\pi}{r} \sum_{j \in \mathcal{V}_i} \sin\left(\frac{\pi}{r} d^j\right) (\cos(\theta)\cos(\theta_j) + \sin(\theta)\sin(\theta_j)) \\
&= \frac{\pi}{r} \sum_{j \in \mathcal{V}_i} \phi_j \cos(\theta) + \psi_j \sin(\theta) \\
&= \Phi \cos(\theta) + \Psi \sin(\theta) \\
&= \sqrt{\Phi^2 + \Psi^2} \cos(\theta - \vartheta),
\end{aligned} \tag{3.31}$$

where $\phi_j = \sin(\frac{\pi}{r} d^j) \cos(\theta_j)$, $\psi_j = \sin(\frac{\pi}{r} d^j) \sin(\theta_j)$, $\Phi = \frac{\pi}{r} \sum_{j \in \mathcal{V}_i} \phi_j$, and $\Psi = \frac{\pi}{r} \sum_{j \in \mathcal{V}_i} \psi_j$. If $\Phi \neq 0$, $\vartheta = \tan^{-1}(\frac{\Psi}{\Phi}) + m\pi$; otherwise, $\vartheta = sign(\Psi)\frac{\pi}{2} + m\pi$, where $sign(\Psi) = 1$, if $\Psi > 0$; $sign(\Psi) = -1$, otherwise. Notice that the value of integer m depends on the signs of Φ and Ψ, and should make $\vartheta \in [0, 2\pi)$ hold.

So far, how UAVs should move in the plane (x, y) is determined by the solution to a simple one variable equation (3.31). If the UAV i is in the case of (i) of **D2**, i.e. the possible directions $\theta \in [0, 2\pi)$, one can obtain the solution $\theta_0 = \vartheta$; otherwise, in the case of (ii) or (iii) of **D2**, UAV i needs to choose θ_0 from the candidate sectors to maximize (3.31), which is also easy to solve. Thus, the control input applied to the UAV i is $u = [\cos(\theta_0) \ \sin(\theta_0)]'$.

Finally, the complexity of the proposed algorithm is analyzed. At each UAV, the proposed algorithm repeats every a certain period of time and such a period is consistent with that of broadcasting its current position. In each period, it receives at most $n - 1$ messages including the positions of other UAVs. The complexity is $O(n)$. Then, it computes the sets of $\mathcal{P}_h$, $\mathcal{Q}_h$, and $\hat{\mathcal{P}}_h$. Correspondingly, the complexities are $O(n)$, $O(k)$ and $O(n)$ in the worst case. When the UAV is in the case of (i) in **D2**, it can find the optimal moving direction analytically by solving (3.31); while when it is in the cases of either (ii) or (iii) in **D2**, it needs to check m possible angles (if the maximum turning angle of the UAV is $\frac{2\pi}{m}$). Then, the complexity is either $O(1)$ or $O(m)$. When the UAV is in **D3**, to determine the vertical moving direction, it only needs to evaluate the sign of (3.29). Then, the complexity is $O(1)$. Hence, the overall computational complexity for one UAV is either $O(3n + k + 1)$ or $O(3n + k + m)$.

Although the computational complexity for one UAV is linearly related to the number of UAVs in the network, it is worth pointing out that such complexity is for the worst case, where a UAV receives the broadcast from all the other UAVs. In practice, the UAVs that are too far away from a UAV will not impact its movement decision; while only those which are connected with this UAV in the communication graph G, and those which are within d_{safe} influence its future movement. In other words, a UAV makes movement decision only based on the measured locations of targets and some nearby UAVs' positions no matter how many UAVs the network has. In practice, a threshold can be applied, such that if the received signal at a UAV is below this threshold, it will be ignored.

3.6.3 Evaluation

We consider an area with the size of 800 m by 600 m as shown in Figure 3.22. In this area, there are four GNs located at the corners (marked by black squares). There are totally 15 depots, each of which stores one UAV (marked by circles). The topology of a given communication graph G is also shown in Figure 3.22, where the UAV-GN link is shown by the dash lines and the UAV-UAV link is shown by the solid lines. The UAV deployment region D is an ellipse represented by the thick solid line and the area has two holes shaded by the solid thin lines where UAVs cannot fly. The system parameters are specified as follows: $Z_{min} = 50$ m and $Z_{max} = 250$ m, $R_1 = 400$ m, $R_2 = 350$ m, $d_{safe} = 30$ m, $\alpha = \frac{\pi}{2}$ and every movement in both horizontal and vertical directions is 1 m.

First, the simulation result with 15 UAVs is demonstrated. There are 200 targets randomly deployed in the area and they may be outside D, see Figure 3.23a.

Figure 3.22 The area of interest.

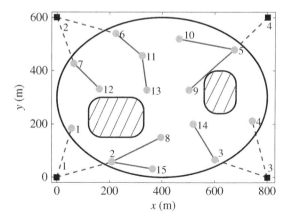

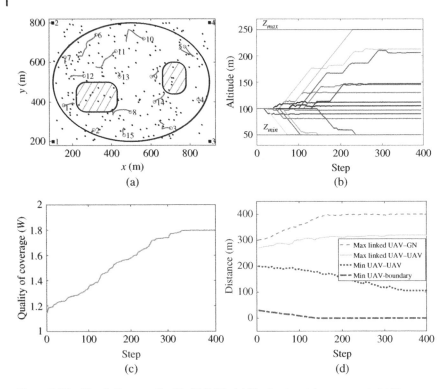

Figure 3.23 Simulation result with 15 UAVs. (a) The horizontal movement. (b) The vertical movement. (c) The QoC. (d) Maximum linked UAV-GN distance, maximum linked UAV-UAV distance, minimum UAV-UAV distance, and minimum UAV-boundary distance.

The initial positions of the UAVs are the locations of the depots as shown in Figure 3.22, and the initial altitudes are all 100 m. Figure 3.23a shows how these 15 UAVs move (indicated by the solid lines connecting a pair of solid circles). It can be seen the UAVs neither enter the holes nor cross the boundary of D. The altitudes of the UAVs during the movements are shown in Figure 3.23b. The QoC of the network is shown in Figure 3.23c. The QoC keeps increasing with step and the network converges in about 360 steps. The final positions of UAVs increase the QoC by 58% compared to the initial positions. Furthermore, Figure 3.23d records several distance metrics during the movement. It is clear that the maximum distance between any pair of linked UAVs does not exceed R_1, the maximum distance between any pair of linked UAV and GN does not exceed R_2, the minimum distance between any pair of UAVs is always larger than d_{safe}, and the minimum distance from UAVs to boundary D is always no smaller than 0.

Furthermore, we demonstrate the simulation result for the above case, but the targets are mobile. In particular, each target randomly selects one moving

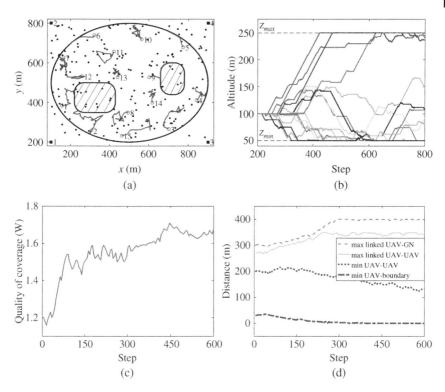

Figure 3.24 Simulation results with 15 UAVs for the case with mobile targets. (a) UAVs' trajectories. (b) UAVs' altitudes. (c) The QoC. (d) Maximum linked UAV-GN distance, maximum linked UAV-UAV distance, minimum UAV-UAV distance, and minimum UAV-boundary distance.

direction and when it meets the boundary of the area, it changes a direction rather than leaving the area. Any movement is bounded by 0.5 m. The simulation result for this case is shown in Figure 3.24. Different from the situation with static targets, the positions of UAVs may not converge to some positions. Thus, we consider running the simulation for 600 steps. Besides, the QoC is not guaranteed to increase all the time, see Figure 3.24c. The reason lies in the movement of targets. However, the proposed algorithm is able to lead the UAVs to some positions where better QoC can be achieved. Similar to the above case, all the links are guaranteed, any pair of UAVs keep away from each other, and no UAVs leave the deployment region $\mathcal{D}$.

To better assess the performance of the proposed approach, a baseline method in [42] is considered for comparison. In [42], UAVs are deployed at the same altitude, and each UAV moves toward the center of mass of its covered targets. Here, the altitude of UAVs is set as the middle of the allowed deployed space, i.e. 150 m. The simulation results for both the static targets and mobile targets are

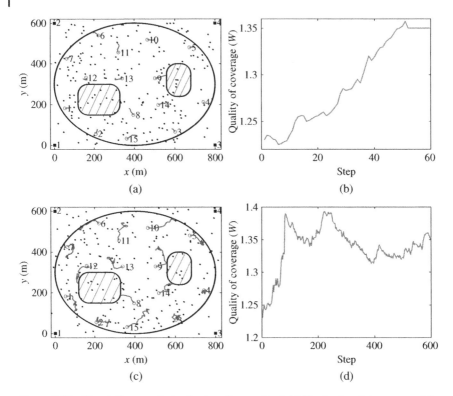

Figure 3.25 Simulation results by the baseline method. (a) The horizontal movement for the static target case. (b) The QoC for the mobile target case. (c) The horizontal movement for the static target case. (d) The QoC for the mobile target case.

shown in Figure 3.25, where the movements of UAVs are shown in Figure 3.25a, c, and the QoCs are shown in Figure 3.25b, d. For the static case, although the compared method converges faster, the proposed approach outperforms of the compared one by about 35% in terms of the QoC. Since the compared method is not aiming at improving QoC, instead its objective is to reduce the average UAV-target distance, the resulting QoC does not increase with step like the proposed algorithm shown in Figure 3.23c. For the mobile case, the proposed algorithm outperforms the compared one by about 25% in QoC.

Moreover, the proposed algorithm and that of [42] are also applied to the cases with other numbers of UAVs. In each of these cases, the UAVs' initial positions are selected from the depots as shown in Figure 3.22. Also, the connectivity requirement should be satisfied at any UAV. In other words, if one depot is selected, the parent depot should be selected as well. For instance, the depot with the index of 13 can only be used if and only if the one indexed by 11 is selected. For a certain

Figure 3.26 Average final QoC in cases with static targets and average approximate final QoC in cases with mobile targets.

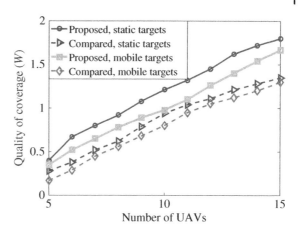

number of UAVs, there are some different and feasible topologies. We randomly pick 10 topologies to conduct the below simulations and the results shown below are the best one among these 10 topologies. We mainly pay attention to the metrics of the final QoC in the cases with static targets, and the approximate final QoC in the cases with mobile targets. The second metric is the average QoC of last 300 steps out of the overall 600 steps. In the cases with static targets, the targets are randomly deployed. To mitigate the influence of the randomness, 10 simulations are conducted and the final QoC in each case is the averaged value. In the cases with mobile targets, the mobility model is consistent with the above case. Also 10 simulations are run, and the approximate final QoC in each case is the averaged value.

The number of UAVs is set from 5 to 15 and the comparisons are displayed in Figure 3.26. It is clear that the proposed algorithm achieves better final QoC in the cases with static targets and approximate final QoC in the cases with mobile targets, and in both situations, the former outperforms the latter by about 20–35%. Also we can see that with the increase of the number of UAVs, the gain achieved by the proposed algorithm against the compared tends to be larger. In other words, for the scenarios where more UAVs are available, the proposed method provides much better QoC.

3.7 Summary and Future Research

In this chapter, we discussed several recent approaches to the deployment UAVs for the surveillance of ground areas and targets. For the former, we first present a triangulation-based deployment, which ensures the asymptotically optimal coverage of a flat area. We then present a 3D Art Gallery-based deployment method, to cover an uneven terrain with buildings that can block the sight of

UAVs. For the latter, we first present a deployment method that enables UAVs to find the locally optimal positions on a given plane. We then present a deployment method that enables UAVs to find the locally optimal positions in the 3D space. While the former two approaches are centralized, which are suitable to predeployment cases, the latter two approaches allow UAVs to find their own positions in a decentralized manner. We provided theoretical analysis of the performance of the developed methods. Moreover, we also presented computer simulations to demonstrate their effectiveness. For the ground areas and targets considered in this chapter, a sufficient number of UAVs will stay over the area of interest. In Chapter 4, we will further investigate the surveillance problems where the number of UAVs is not enough to cover the whole area. We will also consider the cases where the areas and targets of interest are moving.

There are several directions worth being considered further. First, the coverage model used in the approaches discussed in this chapter is a disk model. The good properties of this model lead to the presented analytical solutions. An interesting extension is to consider other types of coverage models for different types of cameras. Second, considering the fact that commercial UAVs are now usually powered by batteries with limited capacity, combining some energy consumption model with the proposed approaches is a promising way to improve the UAV network operation and an important direction for future work.

References

1 C. Luo, J. Nightingale, E. Asemota, and C. Grecos, "A UAV-cloud system for disaster sensing applications," in *2015 IEEE 81st Vehicular Technology Conference (VTC Spring)*, pp. 1–5, IEEE, 2015.

2 C. Yuan, Z. Liu, and Y. Zhang, "UAV-based forest fire detection and tracking using image processing techniques," in *2015 International Conference on Unmanned Aircraft Systems (ICUAS)*, pp. 639–643, IEEE, 2015.

3 M. Rahnemoonfar, R. Murphy, M. V. Miquel, D. Dobbs, and A. Adams, "Flooded area detection from UAV images based on densely connected recurrent neural networks," in *IEEE International Geoscience and Remote Sensing Symposium*, pp. 1788–1791, 2018.

4 A. W. N. Ibrahim, P. W. Ching, G. G. Seet, W. M. Lau, and W. Czajewski, "Moving objects detection and tracking framework for UAV-based surveillance," in *The 4th Pacific-Rim Symposium on Image and Video Technology*, pp. 456–461, IEEE, 2010.

5 A. V. Savkin and H. Huang, "Asymptotically optimal deployment of drones for surveillance and monitoring," *Sensors*, vol. 19, no. 9, p. 2068, 2019.

6 A. V. Savkin and H. Huang, "Proactive deployment of aerial drones for coverage over very uneven terrains: a version of the 3D art gallery problem," *Sensors*, vol. 19, no. 6, p. 1438, 2019.

7 A. V. Savkin and H. Huang, "A method for optimized deployment of a network of surveillance aerial drones," *IEEE Systems Journal*, vol. 13, no. 4, pp. 4474–4477, 2019.

8 H. Huang and A. V. Savkin, "An algorithm of reactive collision free 3D deployment of networked unmanned aerial vehicles for surveillance and monitoring," *IEEE Transactions on Industrial Informatics*, vol. 16, no. 1, pp. 132–140, 2020.

9 R. Kershner, "The number of circles covering a set," *American Journal of Mathematics*, vol. 61, no. 3, pp. 665–671, 1939.

10 M. Jakob, E. Semsch, D. Pavlıcek, and M. Pěchoucek, "Occlusion-aware multi-UAV surveillance of multiple urban areas," in *The 6th Workshop on Agents in Traffic and Transportation (ATT 2010)*, pp. 59–66, Citeseer, 2010.

11 I. Uluturk, I. Uysal, and K.-C. Chen, "Efficient 3D placement of access points in an aerial wireless network," in *2019 16th IEEE Annual Consumer Communications & Networking Conference (CCNC)*, pp. 1–7, IEEE, 2019.

12 S. Fisk, "A short proof of Chvátal's watchman theorem," *Journal of Combinatorial Theory, Series B*, vol. 24, no. 3, p. 374, 1978.

13 I. Mahjri, A. Dhraief, A. Belghith, and A. S. AlMogren, "SLIDE: A straight line conflict detection and alerting algorithm for multiple unmanned aerial vehicles," *IEEE Transactions on Mobile Computing*, vol. 17, no. 5, pp. 1190–1203, 2018.

14 A. Trotta, M. Di Felice, F. Montori, K. R. Chowdhury, and L. Bononi, "Joint coverage, connectivity, and charging strategies for distributed UAV networks," *IEEE Transactions on Robotics*, vol. 34, no. 4, pp. 883–900, 2018.

15 C. Caillouet and T. Razafindralambo, "Efficient deployment of connected unmanned aerial vehicles for optimal target coverage," in *Global Information Infrastructure and Networking Symposium (GIIS)*, pp. 1–8, IEEE, 2017.

16 C. Caillouet, F. Giroire, and T. Razafindralambo, "Optimization of mobile sensor coverage with UAVs," in *Conference on Computer Communications Workshops (INFOCOM WKSHPS)*, pp. 622–627, IEEE, 2018.

17 M. Erdelj, E. Natalizio, K. R. Chowdhury, and I. F. Akyildiz, "Help from the sky: leveraging UAVs for disaster management," *IEEE Pervasive Computing*, vol. 16, pp. 24–32, 2017.

18 J. Araujo, P. Sujit, and J. B. Sousa, "Multiple UAV area decomposition and coverage," in *IEEE Symposium on Computational Intelligence for Security and Defense Applications (CISDA)*, pp. 30–37, IEEE, 2013.

19 L. D. P. Pugliese, F. Guerriero, D. Zorbas, and T. Razafindralambo, "Modelling the mobile target covering problem using flying drones," *Optimization Letters*, vol. 10, no. 5, pp. 1021–1052, 2016.

20 D. Zorbas, L. D. P. Pugliese, T. Razafindralambo, and F. Guerriero, "Optimal drone placement and cost-efficient target coverage," *Journal of Network and Computer Applications*, vol. 75, pp. 16–31, 2016.

21 H. Shakhatreh, A. Khreishah, J. Chakareski, H. B. Salameh, and I. Khalil, "On the continuous coverage problem for a swarm of UAVs," in *the 37th Sarnoff Symposium*, pp. 130–135, IEEE, 2016.

22 E. Semsch, M. Jakob, D. Pavlicek, and M. Pechoucek, "Autonomous UAV surveillance in complex urban environments," in *IEEE/WIC/ACM International Joint Conference on Web Intelligence and Intelligent Agent Technology-Volume 02*, pp. 82–85, IEEE Computer Society, 2009.

23 L. Geng, Y. Zhang, J. Wang, J. Y. Fuh, and S. Teo, "Mission planning of autonomous UAVs for urban surveillance with evolutionary algorithms," in *the 10th IEEE International Conference on Control and Automation (ICCA)*, pp. 828–833, IEEE, 2013.

24 J. O'Rourke, *Art gallery theorems and algorithms*, vol. 57. Oxford: Oxford University Press, 1987.

25 A. Bottino and A. Laurentini, "A nearly optimal algorithm for covering the interior of an art gallery," *Pattern Recognition*, vol. 44, no. 5, pp. 1048–1056, 2011.

26 S. P. Fekete, S. Friedrichs, A. Kröller, and C. Schmidt, "Facets for art gallery problems," *Algorithmica*, vol. 73, no. 2, pp. 411–440, 2015.

27 V. Chvatal, "A combinatorial theorem in plane geometry," *Journal of Combinatorial Theory, Series B*, vol. 18, no. 1, pp. 39–41, 1975.

28 M. Marengoni, B. A. Draper, A. Hanson, and R. Sitaraman, "A system to place observers on a polyhedral terrain in polynomial time," *Image and Vision Computing*, vol. 18, no. 10, pp. 773–780, 2000.

29 P. Theodorakopoulos and S. Lacroix, "A strategy for tracking a ground target with a UAV," in *IEEE/RSJ International Conference on Intelligent Robots and Systems*, pp. 1254–1259, September 2008.

30 S. A. P. Quintero, F. Papi, D. J. Klein, L. Chisci, and J. P. Hespanha, "Optimal UAV coordination for target tracking using dynamic programming," in *the 49th IEEE Conference on Decision and Control (CDC)*, pp. 4541–4546, December 2010.

31 D. Zorbas, T. Razafindralambo, D. P. P. Luigi, and F. Guerriero, "Energy efficient mobile target tracking using flying drones," *Procedia Computer Science*, vol. 19, pp. 80–87, 2013.

32 H. Zhao, H. Wang, W. Wu, and J. Wei, "Deployment algorithms for UAV airborne networks towards on-demand coverage," *IEEE Journal on Selected Areas in Communications*, vol. 36, no. 9, pp. 2015–2031, 2018.

33 S. Sabino and A. Grilo, "Topology control of unmanned aerial vehicle (UAV) mesh networks: a multi-objective evolutionary algorithm approach," in *the 4th*

ACM Workshop on Micro Aerial Vehicle Networks, Systems, and Applications, pp. 45–50, ACM, 2018.

34 N. L. Carothers, *Real analysis.* Cambridge University Press, 2000.

35 P. Ješke, Š. Klouček, and M. Saska, "Autonomous compact monitoring of large areas using micro aerial vehicles with limited sensory information and computational resources," in *International Conference on Modelling and Simulation for Autonomous Systems,* pp. 158–171, Springer, 2018.

36 M. Saska, V. Krátký, V. Spurný, and T. Báča, "Documentation of dark areas of large historical buildings by a formation of unmanned aerial vehicles using model predictive control," in *2017 22nd IEEE International Conference on Emerging Technologies and Factory Automation (ETFA),* pp. 1–8, IEEE, 2017.

37 S. Boyd and L. Vandenberghe, *Convex optimization.* Cambridge university press, 2004.

38 A. Okabe, B. Boots, K. Sugihara, and S. N. Chiu, *Spatial tessellations: concepts and applications of Voronoi diagrams,* vol. 501. John Wiley & Sons, 2009.

39 A. V. Savkin, A. S. Matveev, M. Hoy, and C. Wang, *Safe robot navigation among moving and steady obstacles.* Elsevier, 2015.

40 M. Hoy, A. S. Matveev, and A. V. Savkin, "Algorithms for collision-free navigation of mobile robots in complex cluttered environments: a survey," *Robotica,* vol. 33, no. 3, pp. 463–497, 2015.

41 C. Wang, A. V. Savkin, and M. Garratt, "A strategy for safe 3D navigation of non-holonomic robots among moving obstacles," *Robotica,* vol. 36, no. 2, pp. 275–297, 2018.

42 A. V. Savkin and H. Huang, "Deployment of unmanned aerial vehicle base stations for optimal quality of coverage," *IEEE Wireless Communications Letters,* vol. 8, no. 1, pp. 321–324, 2019.

43 A. V. Savkin and C. Wang, "A simple biologically inspired algorithm for collision-free navigation of a unicycle-like robot in dynamic environments with moving obstacles," *Robotica,* vol. 31, no. 6, pp. 993–1001, 2013.

44 T. R. Jensen and B. Toft, *Graph coloring problems.* John Wiley & Sons, 2011.

45 A. N. Kolmogorov and S. V. Fomin, *Introductory real analysis.* New York: Dover, 1975.

46 H. X. Pham, H. M. La, D. Feil-Seifer, and M. C. Deans, "A distributed control framework of multiple unmanned aerial vehicles for dynamic wildfire tracking," *IEEE Transactions on Systems, Man, and Cybernetics: Systems,* no. 99, pp. 1–12, 2018.

47 P. Basu, J. Redi, and V. Shurbanov, "Coordinated flocking of UAVs for improved connectivity of mobile ground nodes," in *IEEE MILCOM 2004. Military Communications Conference, 2004,* vol. 3, pp. 1628–1634, IEEE, 2004.

48 A. N. Brintaki and I. K. Nikolos, "Coordinated UAV path planning using differential evolution," *Operational Research,* vol. 5, no. 3, pp. 487–502, 2005.

49 B. Zhang, Z. Mao, W. Liu, and J. Liu, "Geometric reinforcement learning for path planning of UAVs," *Journal of Intelligent and Robotic Systems*, vol. 77, no. 2, pp. 391–409, 2015.

50 G. Regula and B. Lantos, "Formation control of a large group of UAVs with safe path planning," in *21st Mediterranean Conference on Control and Automation*, pp. 987–993, IEEE, 2013.

4

Autonomous Navigation of UAVs for Surveillance of Ground Areas and Targets

4.1 Introduction

The application of unmanned aerial vehicles (UAVs) for surveillance and monitoring is an active area of research [1–6]. Due to their relatively low cost, flexibility, maneuverability, rapid navigation capability, and ease of use, UAVs can be efficiently deployed to monitor areas of interest such as a field of plants, border areas, and spreading disaster areas in which UAVs provide valuable information [7–9]. Different from Chapter 3 where a large number of UAVs are used to achieve a persistent coverage, this chapter discusses the usage of UAVs in several surveillance scenarios where persistent coverage cannot be achieved by a given number of UAVs. So, the UAVs need to carry out a periodical surveillance. We start from the scenario to survey an area of interest, which can be for border control and agriculture. We present an asymptotically optimal path planning method for UAVs to monitor the area. We address the revisit period minimization problem and develop an algorithm for navigating a team of UAVs so that every point of a given ground region is periodically surveyed. The result of this manuscript reminds in spirit the main result of [10] where a method for static deployment of UAVs over a ground region was proposed. We prove that the proposed algorithm is asymptotically optimal in the sense that the ratio of the revisit period of the algorithm and the minimum possible revisit period converges to 1 as the area of the ground region tends to infinity. Moreover, it is a construction method requiring a very low computation load.

One of the most dangerous disasters is a bushfire. Every year bushfires destroy millions of acres of land and hundreds of millions of dollars are spent to extinguish these fires. To mitigate this damage, it is necessary to continuously monitor the growth of fires. One of the main challenges is that the spread of bushfires can be very fast, so there is an urgent need to precisely monitor the growth of the bushfire and provide, in real-time, the current coordinates of the frontier of the moving

Autonomous Navigation and Deployment of UAVs for Communication, Surveillance and Delivery,
First Edition. Hailong Huang, Andrey V. Savkin, and Chao Huang.

bushfire region. Since satellite coverage is periodic with unfavorable revisit times, UAV coverage that is available on-demand is preferable. Moreover, bushfire fronts might be very long, therefore, a single UAV would not be able to observe such a large fire boundary even at large altitudes. Therefore, a network of UAVs should be needed. Other spreading disaster processes such as offshore oil spills, flood, and coal ash spills pose similar challenges [1, 2]. Motivated by the necessity of having an effective surveillance approach for spreading disaster processes (including but not limited to bushfire, offshore oil spills, flood, and coal ash spill [1, 2]), we consider a network of UAVs monitoring a moving disaster area. Initially, we aim to navigate the network of UAVs to the frontier of the moving disaster area so that after a finite time the UAVs will always be on the boundary of the moving disaster area. Then, we introduce some measures describing how fast the disaster area is spreading at each particular point of the boundary. Our goal is to navigate the UAVs to locations that maximize some integral of the introduced measure over all points of the boundary of the disaster area that are inside of the visibility cone of at least one UAV. In other words, we want to navigate the UAVs so that they monitor a spreading part of the boundary of the disaster area. We develop a computationally efficient and easily implementable sliding mode control law for UAV navigation. Unlike many other navigation algorithms proposed in the area (see e.g. [1–4, 6]), the developed navigation law achieves the global maximum in the stated optimization problem.

In addition to the monitoring of areas of interest, this chapter also considers traffic surveillance, which has become a significant task in recent years due to the tremendous increase of the number of vehicles [11]. In this application, collecting traffic information has been a key mission because the traffic information plays an important role in road network management such as optimizing journeys for road users. Besides the roadside units (RSUs) and the vehicle-amounted devices, another approach to collecting road traffic information is to explore UAVs [12]. When used in road traffic monitoring, UAVs are promising to bring several advantages. A UAV can fly over a certain part of the road and see a large area with a very low probability of being blocked. Moreover, technological advances make aerial surveillance by UAVs an exponentially cheaper and less inconspicuous option than the conventional surveillance by police helicopters in terms of manufacturing and maintenance costs and the size of the aerial vehicles [13]. Indeed, the lower cost and the smaller size promotes the large-scale application of aerial road traffic monitoring. We consider the scenario of using UAVs to monitor a specific road segment where accidents frequently happen and locate the accidents. When an accident or congestion occurs at a certain position, the traffic speed reduces and the road may be blocked. For example, when a vehicle accident (breakdown or collision) occurs on a one-lane road, the vehicles behind are blocked. Also, some streets may be thronged with people, after sports games, theater, etc. We assume that the UAVs are embedded with the basic image or video processing techniques to provide the necessary sensory data for the control of the UAVs. We propose a decentralized scheme to navigate the UAVs toward the congested road part and enable the UAVs to have a better view of the vehicles. In this scheme, the UAVs can have several modes: *initial, searching, accumulating,* and *monitoring.* In the *initial* mode, the UAVs start from their initial positions and move onto the target road. Then, they start to patrol the road to search for accidents. Once an accident is detected, the UAVs move to that area and try to find better positions to collect more information about the accident. When the accident disappears, the UAVs turn into the *searching* mode. The UAVs need to communicate with others to exchange their locations and the measured information. Each UAV only needs to communicate with at most two neighbor UAVs. Only two UAVs need to share their measurements, and both of them make it possible to implement the scheme in real time. The proposed method can be applied to not only the collection of road traffic information but also the monitoring of a large number of people going to attend or leaving from some events. The police department may wish to know whether someone is injured so that they can take action in the earliest stage. It can also be used in border patrol, intruder detection, and other applications requiring quick detection and then collecting information from the targets.

Another problem under consideration in this chapter is to use UAVs to monitor a group of moving targets moving along a curvy road, not in any direction on the ground [5, 9]. Real-life applications include surveillance of moving ground vehicles for military, security, and policing purposes as well as ground traffic monitoring. The group of moving targets is modeled by a time-varying density function defined on a segment of the road. The speeds of targets are unknown and time-varying. Moreover, different targets in the group might have different speeds. Different from the work aiming on-road patrolling to detect accidents [14, 15], we focus on the optimal surveillance of moving targets. The technique of Voronoi partition is adopted. Different from the area covered in the context of mobile sensor networks, we investigate the dynamic coverage of moving targets. A problem of optimal navigation is stated where the optimized function is an integral measure of distance from all targets in the group to their closest UAVs. A navigation algorithm is proposed. The algorithm is mostly distributed with minor participation of the central controller. The proposed method is relatively computationally simple and easy to implement in real time. The local optimality of the proposed algorithm is proved, i.e. the trajectories of the UAVs converge to optimal locations that deliver the best possible surveillance of the group of moving targets.

Moreover, we consider a scenario where a UAV team carries out video surveillance on moving targets that may belong to different groups, such as pedestrians or vehicles. They may merge into one large group or separate into two or more small groups. Beyond the fundamental observability, i.e. keeping a target within the field of view (FoV) of a UAV, it is also important to deliver high-quality surveillance by reducing the distances between the UAVs and targets so that the UAVs can accurately capture some necessary features of the targets. One typical scenario is that the police uses UAVs to monitor people going to some sports events or returning home after the events via a public transportation station in urban areas. The police may wish to know whether someone is injured so that they can take action in the earliest stage. People that start from different gathering locations may follow the predefined paths, merge after some time at some intersection of their paths and then move toward their destination. Though these paths can be known to the police in advance, what the police may also need is an online method to navigate the UAVs such that the people can be effectively monitored.

To formulate the considered problem, we use a time-varying density function to model the moving targets whose speeds are unknown and may vary with time. A surveillance problem is stated where the average distance from the targets to their closest UAVs is to be minimized. A navigation algorithm is proposed, which enables each UAV to determine its movement using the measured target density, the shared position information of the nearby UAVs and the overall density from a central station. Such a central station can be another UAV that flies higher enough to have better communication links with the surveillance UAVs. It collects the target density from the surveillance UAVs and then sends the normalized density back to each UAV. It may also relay some position information if the UAV–UAV direct communication is blocked, e.g. by buildings. It may act as an observer and directly observe the targets and send the density information to each surveillance UAV. The proposed navigation algorithm is computationally simple, and it guides the UAVs to move toward positions that deliver good quality surveillance of the moving targets from their current positions. Computer simulations are conducted on a street map to illustrate the effectiveness of the method. The main results of this chapter were originally published in [16–20].

The remainder of the chapter is organized as follows. Section 4.2 discusses some closely relevant recent publications. Section 4.3 presents an asymptotically optimal path planning approach for area coverage. Section 4.4 describes a navigation method for UAVs to conduct surveillance on a moving area. Section 4.5 discusses navigation of UAVs for surveillance of targets moving on a road segment. Section 4.6 studies navigation of UAVs for surveillance of moving targets along a road. Section 4.7 presents a navigation method for UAVs to monitor groups of moving targets. Finally, a summary is given in Section 4.8.

4.2 Related Work

A variety of technologies must be integrated to achieve a good quality of surveillance. On the sensor level, many approaches have been proposed for target detection and tracking. For example, the paper [21] aims at locating moving targets in complex road scenes by applying a background subtraction methodology. The authors of [22] introduce a target detecting and tracking system based on image data collected by a UAV. It uses consecutive frames to generate the target's dynamic information, such as positions and velocities. The paper [23] targets developing a UAV-based target tracking and recognition system with the capabilities of accurate camera positioning, fast image processing, and information fusion. With the sensory information, the coordination, path planning, and motion control of UAVs play crucial roles in guaranteeing the quality of surveillance [24]. For example, the paper [25] presents a navigation method for coordinated standoff tracking of moving target groups.

Previous research work has investigated the problem of repetitively covering an area of interest using ground robots. Most approaches first decompose the free region and then compute a global solution. There are two popular approaches for decomposition: grid-based and cellular-based [26]. In grid-based approaches, the region is represented as a grid map, and then a spanning tree is built for path planning [27]. The main difficulty is the efficiency, which depends on the resolution of the map. The cellular decomposition splits the region into nonoverlapping cells, and the widely used method is the boustrophedon decomposition [28]. This method takes some critical points on the boundary of obstacles for decomposition. Other methods such as Voronoi diagram have also been used for cellular decomposition [29]. The main demerit of these approaches is the obtained cells may be unbalanced, which leads to various revisit times of the cells.

With the ongoing growth of vehicle ownership, many urban roads become congested, especially during rush hours. Collecting traffic information has been a significant task since it plays an important role in road network management such as optimizing journeys for road users. There are several ways to collect traffic information. A conventional method is through the deployed RSUs such as underground sensors and low-altitude cameras [30], which passively provide the point-level traffic information like the number of passing vehicles during a specific period at the sensors' locations. Another method is to use some sensors such as GPS devices that are embedded with vehicles to collect vehicle-level traffic information. This approach requires vehicles to carry such devices, and thus cannot guarantee to provide the whole picture of the traffic network.

When used in road traffic monitoring, UAVs can fly over a certain part of the road and see a large area with a very low probability of being blocked. Image processing techniques together with the road information [31] can detect and estimate targets' states from the camera image, and some other information such as velocity can also be estimated [32]. In addition, it is possible for an accident to occur somewhere on a road that cannot be detected by RSUs because two RSUs are some distance away from each other, and the accident may occur outside the detecting range of the two nearest RSUs. Installing more RSUs can solve this problem, but one drawback is the low utility. In contrast, UAVs can be deployed on-demand, and they can detect accidents actively.

Some strategies have been proposed for road traffic monitoring by UAVs. For example, the paper [14] focuses on the route planning of a single UAV to survey a number of road segments. Considering that the traffic monitoring demand may vary with time, the authors propose a real-time rerouting method. The paper [33] further considers using multiple UAVs for road patrol, and the charging issue has been accounted for in the model. The paper [15] formulates a multi-objective optimization problem for UAVs to visit target roads, where the total distance traveled by the UAVs, customer satisfaction and the number of UAVs are jointly considered. In [14, 15, 33], the road segments are abstracted as targets, and the traveling salesman problem (TSP) and the vehicle routing problem are used to formulate the routing problem. The paper [34] considers the road segments directly without abstracting. The authors focus on the arc routing problem to minimize the total cost for surveying the target roads.

4.3 Asymptotically Optimal Path Planning for Surveillance of Ground Areas

We start from a scenario where the number of UAVs carry out a periodical monitoring of a given area of interest, and the goal is to design the paths of the UAVs so that the revisit time of the target area is minimized. In Section 4.3.1, we present the symbols first and then state the problem of interest. The planning algorithm is presented in Section 4.3.2.

4.3.1 Problem Statement

Let $p \in \mathbf{R}^2$ be the vector of Cartesian coordinates on the ground plane. Moreover, let D be a given bounded, closed and Lebesgue measurable region [35] of the ground plane. There are n UAVs labeled $1, 2, \ldots, n$ that flying over D at a given altitude $a > 0$. We assume that the motion of each UAV is described by the equation

$$\dot{p}_i(t) = v_i(t) \tag{4.1}$$

where $p_i(t)$ is the coordinates of the projection of the current position of the UAV i to the ground plane, and $v_i(t)$ is the current UAV velocity vector. We assume that $\|v_i(t)\| \leq V$, where $V > 0$ is given, and $\| \cdot \|$ denotes the standard Euclidean vector norm. This motion model is suitable for rotary-wing UAVs, and it allows the UAVs to make sharp turns. It enables the UAVs to follow the designed paths, as they may have sharp corners as seen later. Moreover, UAVs are equipped with ground facing video cameras with a given observation angle $0 < \phi < \pi$, which defines the visibility cone of UAVs, so that at time t the UAV i at an altitude of h can only see points P of the ground that are inside of the circle of radius given by

$$R := \tan\left(\frac{\phi}{2}\right)h \tag{4.2}$$

centered at $p_i(t)$.

Let $I(P)$ be the set of all time intervals $[m, M]$ where $0 \leq m < M$, during which the point $P \in D$ is not seen from any of n UAVs. Moreover, let

$$\beta(P) := \sup_{[m,M] \in I(P)} (M - m), \quad \beta(D) := \sup_{P \in D} \beta(P). \tag{4.3}$$

Problem Statement: Our objective is to develop a path planning algorithm for n UAVs satisfying (4.1) that solves the following revisit period optimization problem

$$\beta(D) \rightarrow \min. \tag{4.4}$$

Remark 4.1   It is clear that the revisit period $\beta(D)$ is some measure of the quality of surveillance of the ground region D by n UAVs flying over D, in the sense that the smaller $\beta(D)$, the shorter the maximum time during which some point of D is not seen by any UAV. So our objective is to navigate the UAVs to minimize $\beta(D)$.

4.3.2   Path Planning Algorithm

Definition 4.1   Let l be a straight line on the ground plane. A closed planar set $\mathcal{E}$ is said to be l-convex if the intersection of any straight line parallel to l with $\mathcal{E}$ is either empty or a single point or a closed interval, see Figure 4.1.

Remark 4.2   Any convex set is l-convex for any l.

Assumption 4.1   There exist straight lines $l_1, \ldots, l_n$ such that D can be partitioned into n closed, bounded, and l_i-convex regions $D_i, i = 1, \ldots, n$ with equal areas. Moreover, any D_i has piecewise smooth boundary of length L_i.

Figure 4.1   Illustration of an l-convex closed planar set $\mathcal{E}$.

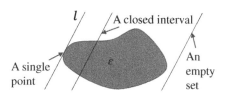

The proposed path planning algorithm for a region D satisfying Assumption 4.1, consists of the following steps.

Step A1: We take l_i and D_i from Assumption 4.1.

Step A2: For any i, we build a sequence of straight lines parallel to l_i at the distance $2r$ between any two closest to each other lines, where R is defined by (4.2), see Figure 4.2.

Step A3: Since D_i is bounded and l_i-convex, the intersection of D_i with the corresponding family of parallel straight lines consists of several parallel segments $S_{i,1}, S_{i,2}, \ldots, S_{i,k(i)}$ (some of them maybe single points). We start at some end of $S_{i,1}$ and move along this segment until we reach the boundary of D_i. Then we move along the boundary until we reach $S_{i,2}$, and then we move along $S_{i,2}$ in the opposite direction until we reach the boundary again, see Figure 4.2.

Step A4: We repeat **Step A3** again and again until we reach the end of $S_{i,k(i)}$, see Figure 4.2.

Step A5: We connect the final point of our path on $S_{i,k(i)}$ with the initial point of our path on $S_{i,1}$ by some curve inside D_i (if possible by a straight line). Since the length of the boundary is L_i, there always exists a path no longer than $\frac{L_i}{2}$.

Step A6: For each i, the UAV i is flying along the closed path constructed in **A1–A5** with the maximum speed V.

Notice that the algorithm **A1–A6** is close to various lawn mowing type path planning algorithms of robotics, see e.g. [36]. We will analyze optimality of the path planning algorithm **A1–A6**. Let $\gamma > 1$ be some number. Introduce the region D_γ obtained from the region D by the linear transformation that maps any point $P \in D$ to the point $(\gamma x, \gamma y)$, see Figure 4.3. So the region D_γ is similar to D but larger. It is also obvious that

$$A(D_\gamma) = \gamma^2 A(D) \tag{4.5}$$

where $A(\cdot)$ denotes the area of a region.

Figure 4.2 Illustration of constructing the closed paths for UAVs.

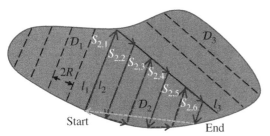

Figure 4.3 Constructing region $\mathcal{D}_\gamma$ from region $\mathcal{D}$.

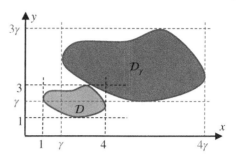

Furthermore, consider some family $\mathcal{F}(\gamma)$ of paths constructed by **Steps A1–A6** over $\mathcal{D}_\gamma$ for all $\gamma > 1$. Let $\hat\beta(\gamma)$ be the value of $\beta(\mathcal{D}_\gamma)$ defined by (4.3) with UAVs flying along the paths $\mathcal{F}(\gamma)$. Let $\beta_0(\gamma)$ be the minimal possible value of $\beta(\mathcal{D}_\gamma)$ for all flying paths of UAVs over the ground region $\mathcal{D}_\gamma$. It is clear that since the region $\mathcal{D}_\gamma$ increases as γ increases, both $\hat\beta(\gamma)$ and $\beta_0(\gamma)$ tend to infinity as γ tends to infinity.

Definition 4.2 A family $\mathcal{F}(\gamma)$ of UAV paths in the regions $\mathcal{D}_\gamma$ is said to be asymptotically optimal, if

$$\lim_{\gamma \to \infty} \frac{\hat\beta(\gamma)}{\beta_0(\gamma)} = 1. \tag{4.6}$$

In other words, path planning is asymptotically optimal, if as the ground region becomes larger, the quality of surveillance with these paths defined by (4.3) becomes close to the best possible quality of surveillance.

Theorem 4.1 Let $\mathcal{F}(\gamma)$ be the family of UAVs paths in the regions $\mathcal{D}_\gamma$ constructed by the path planning algorithm **A1–A6**. Then $\mathcal{F}(\gamma)$ is asymptotically optimal. Moreover,

$$\lim_{\gamma \to \infty} \frac{\hat\beta(\gamma)}{\gamma^2} = \frac{A(\mathcal{D})}{2nVR} \tag{4.7}$$

where $A(\cdot)$ denotes the area of a region.

Proof. First we prove that the family of UAVs paths $\mathcal{F}(\gamma)$ constructed by **A1–A6** satisfies (4.7). Indeed, let $H_i(\gamma)$ be the sum of the lengths of all segments parallel straight lines built in $\mathcal{D}_{i\gamma}$ in **Step A2**. Then it is obvious that

$$\lim_{\gamma \to \infty} \frac{A(\mathcal{D}_{i\gamma})}{2RH_i(\gamma)} = 1. \tag{4.8}$$

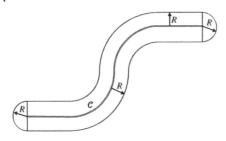

Figure 4.4 *R*-neighborhood of curve *C* (set *S*).

Since $A(D_{i\gamma}) = \gamma^2 A(D_i) = \frac{\gamma^2 A(D)}{n}$, (4.8) implies that

$$\lim_{\gamma \to \infty} \frac{\gamma^2 A(D)}{2nRH_i(\gamma)} = 1. \tag{4.9}$$

Furthermore, it is obvious that the length $T_i(\gamma)$ of the closed UAV trajectory constructed in $D_{i\gamma}$ by **A1–A6** satisfies $H_i(\gamma) < T_i(\gamma) \leq H_i(\gamma) + \frac{3\gamma L_i}{2}$. Since the UAVs are moving with the maximum speed V, this and (4.9) imply (4.7). Now we prove that for any family of paths,

$$\lim_{\gamma \to \infty} \frac{\beta(\gamma)}{\gamma^2} \geq \frac{A(D)}{2nVR} \tag{4.10}$$

Indeed, consider any curve C of length L and the set S of all points at the distance R or less from some point of this curve, see Figure 4.4. Then the area of this set satisfies $A(S) \leq 2LR + \pi r^2$. This implies that the minimum length $F(\gamma)$ of curves in $D(\gamma)$ with the property that for any point $D(\gamma)$ there exists a point of curves at the distance R or less, satisfies

$$\lim_{\gamma \to \infty} \frac{\gamma^2 A(D)}{2RF(\gamma)} \geq 1. \tag{4.11}$$

Since we have n UAVs and their maximum speed is V, (4.11) implies (4.10). This completes the proof of Theorem 4.1. □

4.3.3 Simulation Results

In this section, we show the effectiveness of the proposed approach via computer simulations.

We consider using rotary-wing UAVs to survey a given region D. Here, we do not consider where the UAVs start. Though their starting points impact the time to complete the first trip, they do not influence the revisit time of the later rounds. The first task is to find the set of straight lines satisfying Assumption 4.1. Our method to find this set of lines is stated as follows. Given D, we first find a line l_1' such that D is l_1'-convex and l_1' is tangent to D; see Figure 4.5. Then, we move l_1' in parallel inward D to get another line l_1, such that the sub-region of

Figure 4.5 Constructing a set of straight lines to partition D into n sub-regions with equal areas.

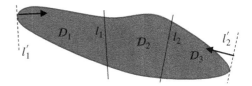

D between l'_1 and l_1, denoted by D_1, satisfies $A(D_1) = \frac{1}{n}A(D)$. Now, we have the first line l_1. Let D_{remain} denote the remaining sub-region after removing D_1 from D, i.e. $D_{remain} = D/D_1$. We then find a line l'_2 such that D_{remain} is l'_2-convex. We move l'_2 in parallel inward D_{remain} to get l_2; see Figure 4.5. We repeat this until D is divided into n sub-regions with equal areas. A simplified version of the above method is stated as follows. After having l_1, instead of finding another line l'_2, we move l_1 further to get l_2, such that (i) l_2 is parallel to l_1 and (ii) the area of the subregion of D between l_1 and l_2 is $\frac{1}{n}A(D)$. By this simplified method, we obtain a set of parallel straight lines. Illustrative examples of the simplified method is shown in Figure 4.6, where $n = 3$, $\alpha = \frac{\pi}{2}$ and $h = 10$ m. We show the cases with $\gamma = 1$ and $\gamma = 2$, respectively. These examples demonstrate how the paths are generated according to the proposed method.

We make a comparison with a benchmark method [37]. This method consists of three steps. First, it partitions the area of interest, which is similar to our method. Then, it grids each sub-area by squares whose edge length is $2R$. The grids overlapping with the sub-area are considered as waypoints. Finally, for each sub-area, it solves a TSP using the Generic Algorithm (GA) to construct a tour that visits each of the waypoints once. We use the MATLAB embedded function to solve the problems. Using the same area as above, the obtained UAVs' trajectories by the benchmark method are shown in Figure 4.7, where $n = 3$, and γ takes 1 and 2, respectively.

We now compare simulation result with Theorem 4.1. Using the same parameters of α and h, we consider different values of γ. As seen from Figure 4.8a, with the

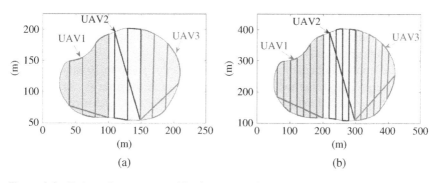

Figure 4.6 Trajectories constructed by the proposed approach. (a) $\gamma = 1$. (b) $\gamma = 2$.

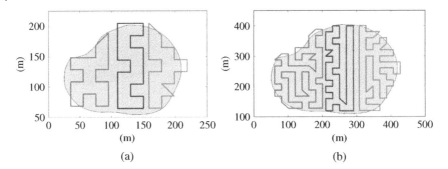

Figure 4.7 Trajectories constructed by the benchmark method. (a) $\gamma = 1$. (b) $\gamma = 2$. Source: Adapted from [38]

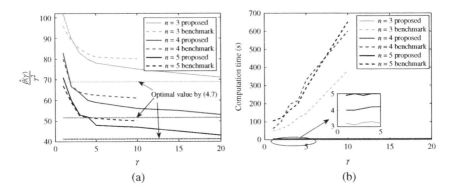

Figure 4.8 Comparison with the benchmark method. (a) $\frac{\hat{\beta}(\gamma)}{\gamma^2}$ versus γ. (b) Computation time versus γ.

increase of γ (from 1 to 30), the value of $\frac{\hat{\beta}(\gamma)}{\gamma^2}$ decreases, and it tends to be the optimal value of $\frac{A(D)}{2nVR}$, where $V = 1$ m/s and $R = 10$ m (computed by (4.2)). Simulation results for other values of n are also shown in Figure 4.8a. We have also applied the proposed method to areas with different shapes, and similar results are obtained. These results are in accordance with Theorem 4.1.

We also apply the benchmark to cases with larger γ (up to 10) and compare it with the proposed method. With the increase of γ, the area $\mathcal{D}_\gamma$ increases, so more grids are needed to fully cover $\mathcal{D}_\gamma$. As solving TSP instances is computationally heavy with respect to the number of points, γ takes up to 10 for the benchmark. As seen from Figure 4.8a, for small values of γ, the benchmark achieves better results. The reason is that in our method when γ is small, the connection between the starting point and ending point, which is constructed according to **Step A5**, takes a relatively large part of a whole path, e.g. see the trajectories of

UAV2 in Figure 4.6. When γ is small, it is easy to optimally solve the TSP instances. For large γ, the proposed method outperforms the benchmark, as it is difficult for GA to search the optimal solution under a given number of iterations. Besides, the proposed method is more computationally efficient. It only takes a few seconds to complete (see Figure 4.8b), while the benchmark can take several minutes. Moreover, the computational time of our method is not sensitive to γ but only to n. Because partitioning the region takes a relatively long time in finding a set of lines $l_1, l_2, \ldots, l_n$, while constructing a trajectory only needs to move a line l_i in parallel to obtain other straight lines. In contrast, the computation time of the benchmark significantly depends on γ, as larger γ means more grids to visit.

4.4 Navigation of UAVs for Surveillance of a Moving Ground Area

Different from Section 4.3 where the target area is static, in this section, we consider the more challenging case where the target area is moving. We firstly present the problem in Section 4.4.1, and then discuss the UAV navigation method in Section 4.4.2.

4.4.1 Problem Statement

A moving disaster area is modeled by time-varying planar regions $D(t)$ where $t \geq 0$. We assume that for any t, $D(t)$ is a bounded connected Lebesgue measurable planar region with a smooth connected boundary $\partial D(t)$ [38].

The moving disaster area is monitored by $n \geq 2$ UAVs labeled $i = 1, 2, \ldots, n$. The aerial UAVs are moving in a plane parallel to the ground plane with a known constant altitude $h > 0$ in continuous time. Let $(x_i(t), y_i(t))$ be the Cartesian coordinates of the UAV i. Also, let $\theta_i(t)$ be the orientation of the UAV with respect to the x-axis, that is $\theta_i(t)$ is measured from the x-axis in the counter-clockwise direction, it takes values in the interval $[0, 2\pi)$ (so the x-axis corresponds to the orientation $\theta_i = 0$). Furthermore, we assume that the UAV i has the linear speed $v_i(t)$ satisfying

$$-V \leq v_i(t) \leq V \tag{4.12}$$

where $V > 0$ is some given maximum speed. Then, the popular kinematic equations of the UAV motion are given by

$$\begin{aligned}
\dot{x}_i(t) &= v_i(t) \cos(\theta_i(t)), \\
\dot{y}_i(t) &= v_i(t) \sin(\theta_i(t)).
\end{aligned} \tag{4.13}$$

The model (4.13) is suitable to describe the motion of various types of vehicles such as multi-rotary UAVs and autonomous ground vehicles; see, e.g. [39–41].

The heading $\theta_i(t)$ and the linear speed $v_i(t)$ are considered as the control inputs of the UAVs (4.13). Moreover, UAVs are equipped with ground facing cameras with a given observation angle $0 < \phi_0 < \pi$, which defines the visibility cone of each UAV, so that at time t a UAV with the coordinates $(x_i(t), y_i(t))$ can see points (x, y, z) of the ground that are inside of the circle of radius R

$$R_0 := \tan\left(\frac{\phi_0}{2}\right)h \tag{4.14}$$

centered at $(x_i(t), y_i(t))$. Moreover, we assume that UAVs have higher quality visibility cones that are defined by an angle $0 < \phi < \phi_0$ and the corresponding radius

$$R := \tan\left(\frac{\phi}{2}\right)h. \tag{4.15}$$

It is obvious that $R < R_0$.

Let $r > 0$ be a given constant such that $R + r < R_0$. Introduce the function $S_r(p, t) \geq 0$ where $p = (x, y) \in \mathbf{R}^2$, $t \geq 0$, and $S_r(p, t)$ is defined as the area of the region consisting of the points of $D(t)$ that are inside of the circle of radius r centered at the point p. Moreover, we assume that $S_r(p, t)$ is differentiable on t and $W_r(p, t) := \frac{dS_r(p,t)}{dt}$.

Definition 4.3 The set $F(t)$ consisting of points $p \in \partial D(t)$ such that $W_r(p, t) > 0$ is called the frontier of the moving disaster area.

It is clear that points p of the disaster area boundary with higher values of $W_r(p, t)$ correspond to quicker moving parts of the frontier of the disaster area, therefore, it is sensible to deploy UAVs to monitor such parts of the frontier.

Let $P_i(t)$ be the point of the boundary $\partial D(t)$ that is closest to UAV i current location $(x_i(t), y_i(t))$, and $d_i(t)$ be the distance between $P_i(t)$ and $(x_i(t), y_i(t))$. Let $\alpha_i(t) \in [0, 2\pi)$ be the direction from $(x_i(t), y_i(t))$ to $P_i(t)$. If $d_i(t) = 0$, i.e. $(x_i(t), y_i(t))$ is on $\partial D(t)$, introduce the direction $\beta_i(t) \in [0, 2\pi)$ that is the direction of the tangent to the boundary $\partial D(t)$ in the counter-clockwise direction, see Figure 4.9a.

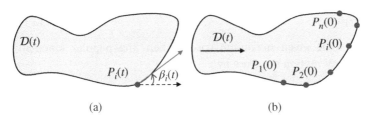

(a) (b)

Figure 4.9 (a) The illustration of the direction of the tangent. (b) The initial closest points.

We assume that the initial closest points $P_1(0), P_2(0), \ldots, P_n(0)$ are located on the frontier $\mathcal{F}(0)$ in this order in the counter-clockwise direction (see Figure 4.9b) and $d_i(0) < R_0$ for all i.

Let $l(P, Q)$ denote the length of the segment of the boundary $\partial D(t)$ between the points P and Q. Furthermore, suppose that all the current UAVs' locations $(x_1(t), y_1(t)), \ldots, (x_n(t), y_n(t))$ belong to the boundary $\partial D(t)$. Then $d_i(t) = 0$ and $P_i(t) = (x_i(t), y_i(t))$ for all $i = 1, \ldots, n$. Introduce the set $\mathcal{V}(t, P_1(t), \ldots, P_n(t)) \subset \partial D(t)$ consisting of all points p of the boundary $\partial D(t)$ such that $l(p, P_i(t)) \le R$ for some i. It is clear that any point p of $\mathcal{V}(t, P_1(t), \ldots, P_n(t))$ is in the higher quality visibility cone of at least one UAV at time t, i.e. there exists i such that p is inside the circle of radius (4.15) centered at $P_i(t)$.

Problem Statement: To develop a navigation law that navigate the UAVs to locations $(x_i(t), y_i(t)) = P_i(t)$ on the boundary $\partial D(t)$ for $i = 1, 2, \ldots, n$ that deliver maximum in the following optimization problem:

$$\int_{\mathcal{V}(t)} W_r(p, t) dp \rightarrow \max. \tag{4.16}$$

In real situations, smoke in the air and high temperature above the fire area may create a big difficulty for the UAVs. One possible approach to these issues is that the boundary of the moving region could be taken not the actual fire boundary but an imaginary line on a certain relatively safe distance from the actual fire boundary.

4.4.2 Navigation Law

Since the cost function (4.16) is a continuous function on a compact set, the global maximum in (4.16) is achieved [38] for all t at some UAV locations $(x_i(t), y_i(t)) = P_i^0(t)$, $i = 1, \ldots, n$. In general, these locations may be non-unique.

Definition 4.4 A navigation law for the UAVs $1, \ldots, n$ is said to be optimal, if there exists a time $t_0 > 0$ such that for all $t \ge t_0$, the UAVs locations $(x_i(t), y_i(t))$ on the boundary $\partial D(t)$ and $(x_i(t), y_i(t)) = P_i^0(t)$ for $i = 1, 2, \ldots, n$ where the locations $P_i^0(t)$ deliver global maximum in (4.16).

Let $\text{sgn}(\cdot)$ denote the standard sign function, i.e.

$$\text{sgn}(z) := \begin{cases} 1, & \text{if } z > 0, \\ 0, & \text{if } z = 0, \\ -1, & \text{if } z < 0. \end{cases} \tag{4.17}$$

Furthermore, let $P_1^-(t)$ denote the point of the boundary of $\partial D(t)$ in the clockwise direction from $P_1(t)$ such that $l(P_1^-(t), P_1(t)) = R$, and $P_n^+(t)$ denotes the point of the boundary of $\partial D(t)$ in the counter-clockwise direction from $P_n(t)$ such that $l(P_n(t), P_n^+(t)) = R$. Introduce the following navigation law:

$$\text{if } d_i(t) > 0 \text{ then } \theta_i(t) = \alpha_i(t), \ v_i(t) = V \tag{4.18}$$

for any $i = 1, \dots, n$,

$$\text{if } d_i(t) = 0 \text{ then}$$

$$\theta_i(t) = \beta_i(t) \ \forall i,$$

$$v_1(t) = \text{sgn}(l(P_1(t), P_2(t)) - 2R)V, \tag{4.19}$$

$$v_n(t) = \text{sgn}(W_r(P_n^+(t)) - W_r(P_1^-(t)))V,$$

$$v_i(t) = \text{sgn}(l(P_i(t), P_{i+1}(t)) - l(P_i(t), P_{i-1}(t)))V,$$

for $i = 2, \dots, n-1$. The proposed navigation law belongs to the class of sliding mode control laws, see e.g. [42]. Unlike classical sliding mode controllers [42], the proposed navigation law is designed to keep the UAVs on a quickly moving line. This gives an advantage over many other methods of UAV navigation that are less capable to react on a quickly changing environment [8, 9] (see the surveys [7, 43] for more details).

4.4.2.1 Available Measurements

Each UAV has the visibility radius R_0. Since $R + r < R_0$, each UAV i is able to calculate the value $W_r(p, t)$ for all p that is no further than R from the UAV's current location. To implement the navigation law (4.18), (4.19), each UAV i needs to know the length of the segments of the boundary from its current location to the neighboring UAVs $i - 1$ and $i + 1$ for $i = 2, \dots, n - 1$, the UAV 2 for $i = 1$ and the UAV $n - 1$ for $i = n$. This is possible if the distance between the UAVs i and $i + 1$ is less than $2R_0$ and these neighboring UAVs can communicate and exchange the measurement information. Also, when a UAV is not on the boundary, it needs to know the coordinates of the closest point of the boundary which is possible as the UAV is initially at a distance less than R_0 to the boundary. Moreover, we assume that there is a central controller that communicates to the UAVs 1 and n so that the UAV n knows $W_r(P_1^-(t))$ which is needed to implement the law (4.19). In practice, the fire may jump from one place to another. In this case, the central controller should be able to send the UAVs to other locations. When the UAVs are again near the fire boundary, they will be navigated by (4.19).

Assumption 4.2 Let p and q be arbitrary points on the plane, and let $P(t)$ and $Q(t)$ be the point of the boundary $\partial D(t)$ that is closest to p and q, correspondingly.

Then, there exist constants $c_1 > 0, c_2 > 0$ and $c_3 > 0$ such that $\|P(t_1) - P(t_2)\| \le c_1(t_2 - t_1)$, $\|l(P(t_1), Q(t_1)) - l(P(t_2), Q(t_2))\| \le c_2(t_2 - t_1)$ and $\|W(P(t_1, t_1) - W(P(t_2, t_2))\| \le c_3(t_2 - t_1)$ for all $t_2 > t_1 \ge 0$, where $\| \cdot \|$ denotes the standard Euclidean distance. Moreover, let $p(t)$ and $q(t)$ be any roots of the equation $W_r(p, t) = W_r(q, t)$. Then there exists a constant $c_4 > 0$ such that $\|p(t_1) - p(t_2)\| \le c_4(t_2 - t_1)$, $\|q(t_1) - q(t_2)\| \le c_4(t_2 - t_1)$ for all $t_2 > t_1 \ge 0$.

Assumption 4.3 For any t, the frontier $F(t)$ of the moving disaster area is a connected segment with the length $l(t)$ such that $2R < l(t) < C$ where $C > 0$ is some constant. Furthermore, for any t, the function $W_r(p, t)$ achieves maximum at some point of the frontier $p^0(t)$ and is monotonically decreasing along the frontier in both clockwise and counter-clockwise directions.

Proposition 4.1 *Suppose the assumptions 4.2 and 4.3 hold with some c_1, c_2, c_3, and c_4 such that*

$$c_1 + c_2 + 2c_3 + 2c_4 < V. \tag{4.20}$$

Then the navigation law (4.18), (4.19) is optimal.

Proof. It obviously follows from (4.18) and Assumption 4.2 that there exits $t_* > 0$ that for all $t \ge t_*$, all n UAVs will be on the boundary $\partial D(t)$. Now for all $t \ge t_*$, introduce the following Lyapunov function (see e.g. [44]):

$$F(t) := \|W_r(P_n^+(t)) - W_r(P_1^-(t))\| + \|l(P_1(t), P_2(t)) - 2R\|$$
$$+ \sum_{i=2}^{n-1} \|l(P_i(t), P_{i+1}(t)) - l(P_i(t), P_{i-1}(t))\|.$$

It is obvious that $F(t) \ge 0$. Furthermore, it follows from (4.18), (4.19), Assumptions 4.2 and (4.20) that $F(t_2) - F(t_1) \le -\epsilon(t_2 - t_1)$ for all $t_2 > t_1 \ge t_*$ where ϵ is any constant such that $0 < \epsilon < V - c_1 - c_2 - 2c_3 - 2c_4$. Therefore, there exists some $t_0 \ge t_*$ such that $F(t) = 0$ for all $t \ge t_0$. Moreover, it is obvious that if $F(t) = 0$ then

$$l(P_i(t), P_{i+1}(t)) = 2R \quad \forall i = 1, \ldots, n-1,$$
$$W_r(P_n^+(t)) = W_r(P_1^-(t)). \tag{4.21}$$

Finally, Assumption 4.3 implies that the global maximum in (4.22) exists and it is achieved at the only set of UAV positions $P_1(t), \ldots, P_n(t)$ satisfying (4.21). This completes the proof of Proposition 4.1. $\qquad \square$

Remark 4.3 As seen in the proof, the optimality of the proposed navigation law does not depend on the initial positions of the UAVs. However, the initial positions impact the process of convergence. Both the convergence time and the trajectory of each UAV are influenced by the initial positions. Also, the inequality (4.20) guarantees that the UAV maximum speed V is larger than the maximum speed of the wild fire spreading. Otherwise, the UAVs are not guaranteed to follow on the frontier of the moving disaster area.

4.4.3 Simulation Results

We first consider the propagation of a bushfire; see Figure 4.10. To get a continuous movement of the bushfire, we use the image processing technique to find out the boundary of each image and interpolate some intermediate boundaries by assuming that a certain point on the boundary moves with a constant speed between two time instants. Some parameters are set as follows: $h = 30$ m, $\phi_0 = \frac{2\pi}{3}$, $\phi = \frac{\pi}{2}$, $R_0 = 52$ m, $R = 30$ m, $r = 15$ m and $V = 3$ m/s. The movements of $n = 5$ UAVs are displayed in Figure 4.11. The right part of the bushfire grows faster. Thus, the boundary of the right part is the so-called frontier. From some initial positions, the UAVs firstly move onto the boundary and then move along the mobile boundary to find the segment that leads to a larger value of (4.16). The objective function values in (4.16) during the movement are shown in Figure 4.12.

We apply the method of [6] to this case. This method regards the boundary of the bushfire as more important than the central area. It is based on an intensity function that sets the fire source with the highest intensity while the boundary with the lowest intensity. The aim is to navigate the UAVs to monitor the boundary. To make it work for our case, we assume that there is one fire source at the center. Starting from the same initial positions, the UAVs' trajectories are shown in Figure 4.11d. The UAVs keep on the boundary of the bushfire. However, they do not look for the segment of the boundary that moves faster. Thus, the objective function values as shown in Figure 4.12 are in general lower than the proposed method.

The above case shows a simple scenario where the mobile disaster area is mainly toward one direction and the boundary is convex. We create a non-convex disaster

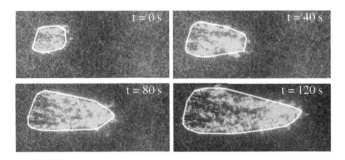

Figure 4.10 The propagation of bushfire.

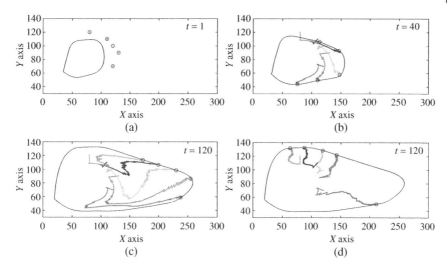

Figure 4.11 (a–c) Monitoring the frontier of bushfire. Source: Modified from https://youtu.be/8Bg1sKfthEA. (d) The movements guided by the method of [6]. Source: Modified from https://youtu.be/hcbp77WN0G8.

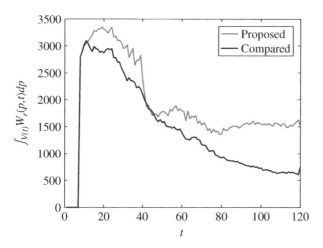

Figure 4.12 Objective function values during the movement.

area as shown in Figure 4.13a. It firstly moves toward the right-top corner, then the right-bottom corner, and finally the x-axis. The movements of 5 UAVs in 300 seconds are shown in Figure 4.13. The UAVs firstly move onto the mobile boundary. Then, they move to the right-top segment, see Figure 4.13b. Furthermore, they move clockwisely to catch the fast-moving part of the boundary, see Figure 4.13c.

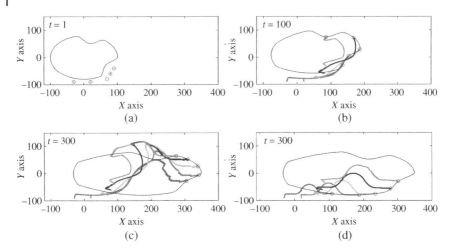

Figure 4.13 (a–c) Monitoring a complex mobile area. Source: Modified from https://youtu.be/K8fsQBGNj20. (d) The movements guided by the method of [6]. Source: Modified from https://youtu.be/cKrzcrEnDDc.

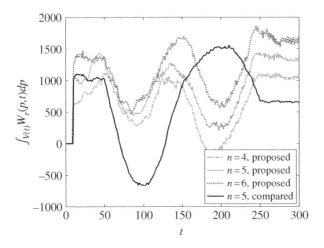

Figure 4.14 Objective function values for the complex scenario.

The objective function values during the movement are shown in Figure 4.14. Simulations with 4 and 6 UAVs are conducted and their objective function values are shown in Figure 4.14. We notice that during a period from about 180 to 220 seconds, the objective function values of the case with 4 UAVs are negative and they are also at low levels for the cases with 5 and 6 UAVs. The reason is that the area suddenly changes the moving direction. After some time, the UAVs adapt their positions to the part of the boundary moving faster.

The method of [6] is also applied to this scenario, see Figure 4.13d. The UAVs on the boundary do not look for the part that moves faster. In the period from 60 to 140 seconds, the UAVs are on a segment having a negative value of $W_r(p, t)$. Although in the period from 150 to 230 seconds the objective function values are larger than those of the proposed method, they are just achieved passively, since the UAVs have not changed their relative positions on the boundary.

4.5 Navigation of UAVs for Surveillance of Moving Targets on a Road Segment

In this section and the following section, we consider the surveillance of targets moving on some road.

4.5.1 Problem Statement

Consider a smooth and curvy road $\mathcal{L}$ whose ending points are A and B, respectively, see Figure 4.15. In general, vehicles can traverse the road $\mathcal{L}$ in two directions.

The road $\mathcal{L}$ is monitored by n UAVs labeled $i = 1, 2, \ldots, n$, where $n > 2$. The UAVs are deployed at an altitude $h > 0$. Let $P_i(t) = (x_i(t), y_i(t))$ be the Cartesian coordinates of the UAV i. Let $\theta_i(t)$ be the orientation of the UAV. It is measured from the x-axis in the counter-clockwise direction, and it takes values in the interval $[0, 2\pi)$ (so the x-axis corresponds to the orientation $\theta_i = 0$). The UAV i has the linear speed $v_i(t)$ satisfying constraint (4.12). The UAV motion is described by Eq. (4.13).

Since a UAV's vision cone can be much larger than the road width, we use the middle separation line to represent the road, see Figure 4.16. Without introducing a new symbol, we still use $\mathcal{L}$ to represent the road. We assume that when the UAVs reach the road, their positions $P_1(0), \ldots, P_i(0), \ldots, P_n(0)$ are located in order on $\mathcal{L}$, see Figure 4.16a. Let $\alpha_i(t) \in [0, 2\pi)$ denote the direction of the tangent to the road $\mathcal{L}$ at $P_i(t)$ in the counter-clockwise direction, see Figure 4.16a.

Let $Y(p) \geq 0$ ($p \in \mathcal{L}$) be the measurement when a UAV is at point p, such as the number of vehicles that are inside of the circle of radius R centered at the point p. We aim at developing a navigation law that navigates the UAVs to locations $P_i(t)$

Figure 4.15 The considered road $\mathcal{L}$.

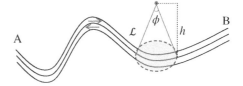

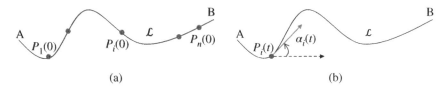

Figure 4.16 (a) The initial closest points. (b) The illustration of the direction of the tangent.

on the road $\mathcal{L}$ for $i = 1, 2, \ldots, n$ such that

$$\sum_{i=1}^{n} Y(P_i(t)) \rightarrow \max. \tag{4.22}$$

We assume that the measurements of the UAVs are additive.

Below, we present the navigation method to achieve (4.22). Let the function $l(P, Q)$ give the length of the segment of $\mathcal{L}$ between the points P and Q.

4.5.2 Proposal Solution

We propose a decentralized scheme to navigate the UAVs toward the congested road part and enable the UAVs to have a better view of the vehicles. The UAVs can have four modes: *initial*, *searching*, *accumulating* and *monitoring*. In the *initial* mode, the UAVs start from their initial positions and move onto the target road. Then, they start to patrol the road to search for accidents. Once an accident is detected, the UAVs move to that area and try to find better positions to collect more information about the accident. When the accident disappears, the UAVs turn into the *searching* mode. Below, we describe these modes in detail.

4.5.2.1 Monitoring Mode

We introduce the following sliding mode navigation law (which is called the *monitoring* navigation law), where the headings and linear speeds of UAVs are set as:

$$\theta_i(t) = \alpha_i(t), \forall i = 1, \ldots, n; \tag{4.23}$$

$$v_1(t) = \begin{cases} V\operatorname{sgn}(l(P_1(t), P_2(t)) - 2R), \text{ if } P_1(t) \neq A, \\ 0, \text{ otherwise.} \end{cases} \tag{4.24}$$

$$v_n(t) = \begin{cases} V\operatorname{sgn}(Y(P_n(t)) - Y(P_1(t))), \text{ if } P_n(t) \neq B, \\ 0, \text{ otherwise.} \end{cases} \tag{4.25}$$

$$v_i(t) = V\operatorname{sgn}(l(P_i(t), P_{i+1}(t)) - l(P_i(t), P_{i-1}(t))), \tag{4.26}$$

for $i = 2, \ldots, n - 1$.

In other words, under the *monitoring* navigation law, each UAV moves along the targeted road segment $\mathcal{L}$, see (4.23). Besides, the two end UAVs, i.e. UAV 1 and UAV n will not cross the ending points A and B, see (4.24) and (4.25), Moreover, the UAVs keep a $2R$ distance away from each other. This not only avoids collision but also decreases the overlap of the vision cone of two UAVs. For any UAV $i = 2, \ldots, n - 1$, it needs to know the positions of UAVs $i - 1$ and $i + 1$, which can be obtained via communication. Together with the knowledge of the road $\mathcal{L}$, both $l(P_i(t), P_{i-1}(t))$ and $l(P_i(t), P_{i+1}(t))$ can be computed. For UAV 1, it only needs to know the position of UAV 2. For UAV n, it needs the information of $Y(P_1(t))$. Clearly, each UAV needs to communicate with at most two UAVs and these UAVs can be predefined based on the positions $P_1(0), \ldots, P_n(0)$. Furthermore, the exchanged information only contains the coordinates and the measured value of $Y(P_1(t))$. Such little overhead makes the navigation law feasible to be implemented in real time. If each UAV is equipped with a sensor to measure the nearby UAVs' positions, the only information to be exchanged is the measured value of $Y(P_1(t))$ from UAV 1 to UAV n, while the middle $n - 2$ UAVs do not need to communicate. It is worth mentioning that by the *monitoring* navigation law (4.23)–(4.26), the UAVs move along the targeted road segment. Although moving off the road could save flight time, moving along the road enables the UAVs to better monitor the vehicles on the road. Moreover, in scenarios where the road is surrounded by mountains, flying off the road may be impossible.

Suppose the curvature $k(p)$ of the road $\mathcal{L}$ satisfies $|k(p)| < \frac{1}{R}$. Then, the navigation law (4.23)–(4.26) guarantees that there is no overlap between the coverage of two nearby UAVs. If $Y(P_i(t))$ measures the number of vehicles, $Y(P_i(t))$ $(i = 1 \ldots, n)$ are additive, and $\sum_{i=1}^{n} Y(P_i(t))$ is the total number of covered vehicles. In addition to vehicles, the method also applies to the surveillance of people, which could indeed be treated as points for a suitable altitude h.

4.5.2.2 Initial Mode

Note that the *monitoring* navigation law (4.23)–(4.26) applies to the situation where the UAVs are already on the road $\mathcal{L}$. In the cases when UAVs start from somewhere off the road, we introduce an *initial* navigation law. Let $z_i \in \mathcal{L}$ denote the point on the road that is the closest to UAV i, i.e. $z_i(t) := \text{argmin} |P_i(t), z|$, where $z \in \mathcal{L}$. Clearly, with the movement of UAV i, $z_i(t)$ varies. Let $\beta(P_i(t), z_i(t))$ denote the direction from $P_i(t)$ pointing to $z_i(t)$. The UAVs should follow the *initial* navigation law:

$$\theta_i(t) = \beta(P_i(t), z_i(t)), \tag{4.27}$$

$$v_i(t) = \begin{cases} |P_i(t), z_i(t)|, & \text{if } |P_i(t), z_i(t)| \leq V, \\ V, & \text{otherwise,} \end{cases} \tag{4.28}$$

for any $i = 1, \ldots, n$. In other words, each UAV moves toward the nearest point on the road in the fastest speed while satisfying constraint (4.12).

With (4.27) and (4.28), the UAVs can move onto the road in the shortest time, and with (4.23), (4.24), (4.25), and (4.26), the UAVs can move to better positions to monitor the ground vehicles.

4.5.2.3 Searching Mode

A demerit of the *monitoring* navigation law (4.23)–(4.26) is that the UAVs may take a long time or even fail to locate the congested road segment. The reason is that such a law is based on the measured values of two ending UAVs, see Eq. (4.25). If the current positions of UAVs are far from the congested road segment, the measured values of the two end UAVs may be very close. It may take a relatively long time for one of the end UAVs to detect the event of interest. In a bad case, the congestion may disappear before being detected.

To this end, a *searching* navigation law is introduced. When the UAVs have not detected any event, they operate in the *searching* mode. To minimize the searching time, we wish that each UAV searches a road segment with the same length. Given that we do not know where events will occur and the UAVs fly at the same speed, the same flying length results in the minimum time to detect an event averagely. Note that the searching of traffic accidents only requires the UAVs to patrol the road, which is different from the target searching strategies such as Levy flight and Brownian search considered in [45]. The road $\mathcal{L}$ can be divided into n segments with equal length. Then, each UAV can move to the corresponding segment and then patrol the segment. Specifically, if UAV i at $P_i(0)$ is outside the corresponding road segment, it first moves to that segment. Then, it moves between the two division points of the segment. When it meets one division point, it changes its heading and moves toward the other division point. During the movement, one UAV may meet its neighbor UAV near the division point. Then, the UAV can make a turn if the distance from the other UAV is less than $2R$. The termination condition of the *searching* mode is that the measurement at some UAV is over a given threshold $\delta_{s \to a}$:

$$\max_{i=1,\ldots,n} Y(P_i(t)) > \delta_{s \to a}. \tag{4.29}$$

The subscript of $\delta_{s \to a}$ means the transfer from the *searching* mode to the *accumulating* mode (explained below). When $Y(p)$ gives the number of the covered targets when a UAV is at p, $\delta_{s \to a}$ represents a threshold number, and such a number can be obtained by observing the real situations of accidents.

4.5.2.4 Accumulating Mode

Suppose it is UAV j that detects the event. Then, all other UAVs set their headings and the speeds as follows:

$$\theta_i(t) = \alpha_i(t), \forall i = 1, \ldots, n, i \neq j; \tag{4.30}$$

$$v_i(t) = \begin{cases} -V, & \text{if } 1 \leq i < j, \ l(P_i(t), P_{i+1}(t)) > 2R, \\ V, & \text{if } j < i \leq n, \ l(P_{i-1}(t), P_i(t)) > 2R, \\ 0, & \text{otherwise.} \end{cases} \tag{4.31}$$

Following (4.30) and (4.31), the UAVs accumulate to the position of UAV j in the speed of V, until the internal distance between two neighbor UAVs becoming $2R$. The termination condition of *accumulating* is that all UAVs have stopped:

$$\max_{i=1,\ldots,n} v_i(t) = 0. \tag{4.32}$$

Then, the UAVs turn into the *monitoring* mode, guided by the *monitoring* navigation law (4.23)–(4.26).

When the event disappears, the UAVs should turn into *searching* mode. We consider the following condition:

$$\max_{i=1,\ldots,n} Y(P_i(t)) < \delta_{m \rightarrow s}, \tag{4.33}$$

where $\delta_{m \rightarrow s}$ is a given threshold indicating that the road condition turns normal, and the subscript means the transfer from the *monitoring* mode to the *searching* mode.

For a brief summary, at the beginning, the UAVs are in the *initial* mode and they follow (4.27) and (4.28) to reach the road. If no event is detected, the UAVs are in the *searching* mode. When the condition (4.29) holds, the UAVs accumulate to the UAV that detects the event following (4.30) and (4.31). When the condition (4.32) holds, the UAVs turn into the *monitoring* mode and they follow (4.23)–(4.26). When the event disappears, i.e. the condition (4.33) holds, the UAVs turn into the *searching* mode. How the UAVs change their working modes is summarized in Figure 4.17.

Remark 4.4 Due to the nature of (4.22) and that the navigation laws are based on sliding mode control, the UAVs may move back and forth frequently. This may lead to some unnecessary movements, which may result in more energy consumption. To address this, we can introduce a dead-zone to avoid the frequent back and

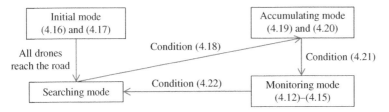

Figure 4.17 The diagram of the proposed navigation scheme.

forth movement, Moreover, the navigation laws are for a planar motion, and the performance depends on the altitudes of UAVs. Practically, due to various factors such as wind, the UAVs may not always remain at the given altitude. To address this issue, we can modify the navigation laws. When two UAVs are at h_1 and h_2, their coverage radii become $R_1 = \tan(\frac{\phi}{2})h_1$ and $R_2 = \tan(\frac{\phi}{2})h_2$, respectively. Then, the term $2R$ in Eq. (4.24) becomes $R_1 + R_2$. As the UAVs can share their location information with their neighbors, UAV 1 can still choose the right moving direction. Other navigation laws are also needed to be modified, and this is our future work.

4.5.3 Simulation Results

We present simulation results to demonstrate the effectiveness of the proposed method. The measurement of each UAV is the number of covered targets. The simulation parameters are $\delta = 8$, $V = 5$ m/s, $\phi = \frac{\pi}{2}$, $h = 30$ m, $\delta_{s \to a} = 8$, $\delta_{m \to s} = 4$, and the average speed of targets is 2 m/s.

We firstly consider a static case (case 1) to demonstrate the convergence of the proposed method. On a given road with two lanes, we randomly place 50 targets, see Figure 4.18a. Four UAVs start from their initial positions and then move onto the road, see Figure 4.18b. In the *searching* mode, one UAV detects the blockage, and then the other UAVs move to that UAV, see Figure 4.18c. Under the *monitoring* navigation law, they reach the final positions, see Figure 4.18d. During the movement, the objective function value, i.e. the number of the covered targets, is shown in Figure 4.19. The result only shows the value after the UAV reaches the road. The proposed navigation laws lead the UAVs to some places to view more targets.

We further consider a mobile case (case 2). Different from case 1, each target is randomly generated and moves at a constant speed with some noise. For two targets generated closely in time, the rare target may move faster than the front one. Any rare target keeps a safe distance from the front one. We simulate two

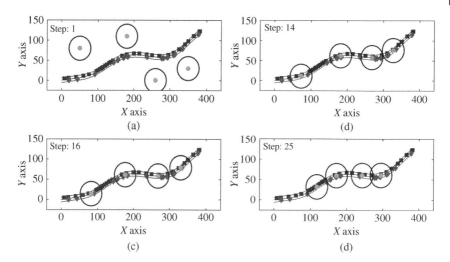

Figure 4.18 Case 1: UAVs' positions at some key steps. The squares represent targets moving right and the diamonds represent targets moving left. The light grey dots are UAVs. The dotted circle is the coverage area of a UAV. UAVs' movements are recorded: source: Modified from https://youtu.be/Ob72y8od2tU.

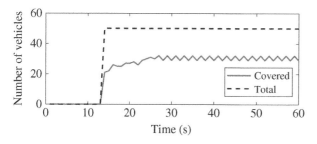

Figure 4.19 Case 1: The covered number of targets against the total number of targets on the road.

blockages by setting a target speed as 0. When a target is too close to the stopped target in front, it stops. The movements of four UAVs are shown in Figure 4.20. From 13 to 204 seconds, the UAVs are in the *searching* mode. At 205 seconds, the first blockage is detected. The UAVs accumulate and monitor the blocked area, see Figure 4.20c, d. At 347 seconds, the UAVs find that the blockage disappears, and they turn into *searching* mode. At 550 seconds, the second blockage is detected. The UAVs accumulate again, see Figure 4.20e, f. During the movement, the

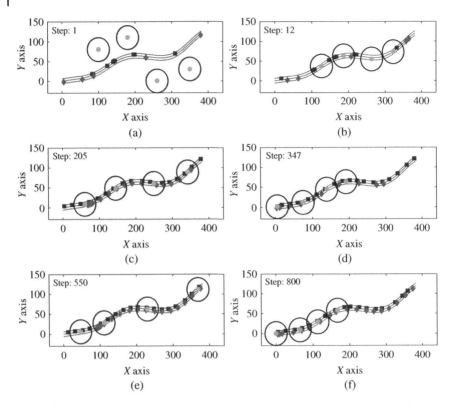

Figure 4.20 Case 2: UAVs' positions at some key steps. The movements are recorded: source: Modified from https://youtu.be/mfxB2qRmOVl.

number of the covered targets is shown in Figure 4.21. These simulation results show that the proposed navigation laws enable UAVs to quickly detect a blockage and then monitor the majority of the targets.

To better demonstrate the performance of the proposed method, we consider a static deployment scheme (like RSUs) and a patrol scheme (as used in [14, 15, 33, 34]) as benchmarks. The static deployment scheme referred to as "static", deploys the same number of UAVs as our method, and the UAVs' positions are selected so that their coverage areas do not overlap. These UAVs passively monitor the passing vehicles. The patrol scheme, referred to as "patrol", sends UAVs to patrol the road segment. For a fair comparison, one UAV is sent every a certain period. Such a period is $\frac{1}{n}$ of the time for a UAV to patrol the road with the speed of V. Figure 4.21 shows that, when no accidents occur, the number of vehicles covered by the proposed method is slightly larger than that by

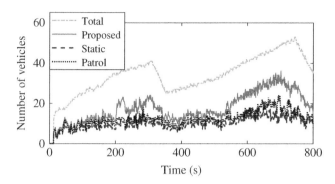

Figure 4.21 Case 2: Comparison of the proposed method and the benchmark methods on the covered numbers of targets.

the benchmark methods. The gap becomes significant when blockages occur. As seen from Figure 4.21, the proposed method can cover much more targets from 200 and 300 seconds, and from 550 to 800 seconds, than the benchmarks.

We present the result of a more realistic case with four lanes in two directions, and the targets can change the lane if (i) a blockage is noticed and (ii) there is a safe gap on the other lane; see Figure 4.22. The proposed method as well as the two benchmarks are applied to this case, and the comparison is shown in

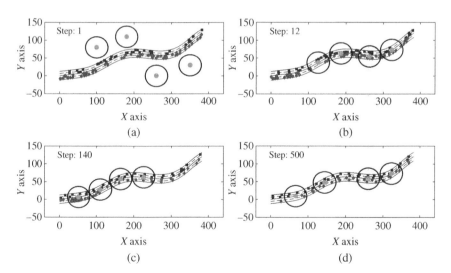

Figure 4.22 Case 3: UAVs' positions at some key steps. The movements are recorded: source: Modified from https://youtu.be/TH1UurzeEh8.

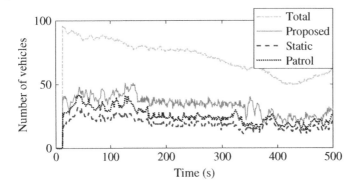

Figure 4.23 Case 3: Comparison of the proposed method and the benchmark methods on the covered numbers of targets.

Figure 4.23. Similar to case 2, the proposed method enables the UAVs to monitor more targets when the accident occurs (between 150 and 350 seconds) than the benchmarks.

4.6 Navigation of UAVs for Surveillance of Moving Targets along a Road

4.6.1 Problem Statement

We assume that a group of ground targets move along a path or road that is modeled by a smooth non-intersecting curve C. We introduce a curvilinear coordinate z along the curve C corresponding to lengths of curve segments, see Figure 4.24. The system is modeled as a discrete-time dynamical system that evolves at discrete time instants $t = 0, 1, 2, \ldots$. The moving group of ground targets at any discrete-time t is modeled by the density function $\rho_t(z)$ that satisfies the following conditions:

$$\rho_t(z) = 0 \quad \forall z > z_I(t);$$

$$\rho_t(z) = 0 \quad \forall z < z_F(t);$$

$$\rho_t(z) > 0 \quad \forall z_I(t) \leq z \leq z_F(t); \qquad (4.34)$$

$$\int_{z_F(t)}^{z_I(t)} \rho_t(z)dz = 1.$$

In other words, it is assumed that the moving group of targets is between the coordinates $z_F(t)$ and $z_I(t)$ at any time t, and the target density is described by the time-varying function $\rho_t(z)$ that is positive between $z_F(t)$ and $z_I(t)$ and zero outside this interval. The frequently used symbols in the chapter are listed in Table 4.1.

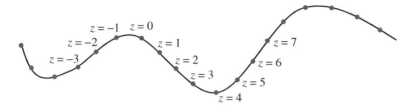

Figure 4.24 A road with curvilinear coordinates.

Table 4.1 Frequently used symbols and their meanings.

Symbol	Meaning
z	Curvilinear coordinate along the considered road
$\rho_t(z)$	Target density function at the coordinate z at time t
h	Flight altitude
$P_i(t)$	Cartesian coordinates of UAV i at t
$U_i(t)$	Displacement of UAV i at t
$d_{\min}(x, y)$	The minimum distance between points x and y
$\phi_0(t)$	Observation angle
R	Radius of the visibility cone on the ground
$C_i(P_1, \ldots, P_n)$	A Voronoi cell of UAV i
$Z_i[t, P_1, \ldots, P_n]$	The gravity center of a Voronoi cell of UAV i at t

The moving group of targets is monitored by n UAVs where $n \geq 2$. All the UAVs are deployed at the same altitude $h > 0$ in one horizontal plane. Hence, we consider the UAVs' Cartesian coordinates at that plane as coordinates on the surface. Let $P_1(t) = (x_1(t), y_1(t)), \ldots, P_n(t) = (x_n(t), y_n(t))$ denote the Cartesian coordinates of the UAVs at time t. Furthermore, the movement of the UAVs is described by the equation:

$$P_i(t + 1) = P_i(t) + U_i(t) \tag{4.35}$$

where $U_i(t)$ is the displacement of the UAV viewed as the control input, and the following constraint holds

$$\|U_i(t)\| \leq U_{\max} \quad \forall i = 1, \ldots, n \tag{4.36}$$

with some $U_{\max} > 0$. Moreover, the UAVs are equipped with ground facing cameras with a given observation angle $0 < \phi_0 < \pi$, which defines the visibility cone of each UAV, so that at time t a UAV with the coordinates $(x_i(t), y_i(t))$ can see points (x, y) of the ground that are inside of the circle of radius R centered at $(x_i(t), y_i(t))$.

Now, let $P_1, P_2, \ldots, P_n$ be some given locations of the UAVs, and introduce the function $d_{\min}(x, y)$ defined for all any point (x, y) in the plane as follows:

$$d_{\min}(x, y) := \min_{i=1,\ldots,n} |(x, y), P_i|, \tag{4.37}$$

i.e. $d_{\min}(x, y)$ is the distance between the point (x, y) and the nearest UAV.

Moreover, for any point of the road C with the curvilinear coordinate z, the planar Cartesian coordinates of this point are denoted by (x_z, y_z). The quality of surveillance delivered by the UAVs placed at time t at the points $P_1(t), P_2(t), \ldots, P_n(t)$ is described by the following function:

$$q(t, P_1(t), \ldots, P_n(t)) := \int_{z_F(t)}^{z_I(t)} d_{\min}^2(x_z, y_z)\rho_t(z)dz. \tag{4.38}$$

In other words, the function $q(t, P_1(t), \ldots, P_n(t))$ is the mathematical expectation $E[d_{\min}^2]$ of the squared distance between a moving target and the nearest UAV at time t. This optimization index and a similar one which is the integral of Euclidean distance $d_{\min}(x_z, y_z)$ have been considered in existing publications [5]. Now we consider the following optimization problem:

$$\min_{P_1(t),\ldots,P_n(t)} q(t, P_1(t), \ldots, P_n(t)) \tag{4.39}$$

where the minimum is taken over all possible UAV locations $P_1(t), \ldots, P_n(t)$.

It is clear that minimization of the function (4.39) is equivalent to maximizing the coverage quality as with a smaller value of the function (4.39), the moving targets are seen from the UAVs from shorter distances.

Definition 4.5 UAV locations $P_1^0(t), \ldots, P_n^0(t)$ is said to be optimal at time t, if

$$d_{\min}(x_z, y_z) \leq R \quad \forall z \in [z_F(t), z_I(t)] \tag{4.40}$$

and $P_1^0(t), \ldots, P_n^0(t)$ deliver the global minimum in the optimization problem (4.39). Correspondingly, $P_1^0(t-1), \ldots, P_n^0(t-1)$ denote UAV locations that deliver the global minimum in the optimization problem (4.39) with time t replaced by $t-1$. Moreover, the trajectories $P_1(t), \ldots, P_n(t)$ of the UAVs (4.35) is said to be locally optimal if

$$\|P_1(t) - P_1^0(t-1)\| \to 0, \ldots, \|P_n(t) - P_n^0(t-1)\| \to 0 \tag{4.41}$$

as $t \to \infty$.

Notice that a UAV navigation algorithm that is optimal in the sense of Definition 4.5 is just optimal one step ahead in time. That is why we call UAV trajectories generated by such an algorithm locally optimal. Such an algorithm may be not optimal over longer time intervals. However, we wish to point out that finding an optimal solution over a longer time interval $[t, t + N]$ would require knowledge of future targets' coordinates over this interval at time t which is not realistic as the targets may have time-varying speeds and the UAVs do not know targets' future coordinates a priori.

Notice that the requirement (4.40) means that any target's possible location is seen by at least one UAV.

Problem Statement: The optimal surveillance problem under consideration is as follows: for the given time-varying target density function $\rho_t(z)$ construct control inputs $U_i(t)$ that generate locally optimal trajectories $P_1(t), \ldots, P_n(t)$ of the UAVs (4.35).

4.6.2 Navigation Algorithm

We will need the following notations. For any $i = 1, 2, \ldots, n$, a Voronoi cell $\mathcal{V}_i(P_1, \ldots, P_n)$ is defined as the set of points $(x_z, y_z) \in C$ of the road such that $|(x_z, y_z), P_i| \leq |(x_z, y_z), P_j|$ for all $j = 1, 2, \ldots, n$, see Figure 4.25. Moreover, the point

$$Z_i[t, P_1, \ldots, P_n] = [X_i(t, P_1, \ldots, P_n), Y_i(t, P_1, \ldots, P_n)]$$

where

$$
\begin{aligned}
X_i(t, P_1, \ldots, P_n) &:= \frac{\int_{\mathcal{V}_i(P_1, \ldots, P_n)} x_z \rho_t(z) dz}{\int_{\mathcal{V}_i(t, P_1, \ldots, P_n)} \rho_t(z) dz}, \\
Y_i(t, P_1, \ldots, P_n) &:= \frac{\int_{\mathcal{V}_i(P_1, \ldots, P_n)} y_z \rho_t(z) dz}{\int_{\mathcal{V}_i(P_1, \ldots, P_n)} \rho_t(z) dz}
\end{aligned}
\tag{4.42}
$$

is called the gravity center of a Voronoi cell $\mathcal{V}_i(P_1, \ldots, P_n)$ at time t. Notice that the gravity center point $Z_i[t, P_1, \ldots, P_n]$ may be and usually is outside the curve C.

Introduce the following navigation algorithm:

if $\|Z_i[t, P_1(t), \ldots, P_n(t)] - P_i(t)\| \leq U_{\max}$

then $U_i(t) := Z_i[t, P_1(t), \ldots, P_n(t)] - P_i(t)$;

otherwise

$$U_i(t) := \frac{U_{\max}[Z_i[t, P_1(t), \ldots, P_n(t)] - P_i(t)]}{\|Z_i[t, P_1(t), \ldots, P_n(t)] - P_i(t)\|} \tag{4.43}$$

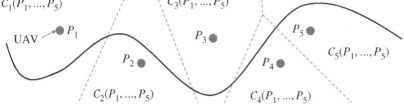

Figure 4.25 An illustration of Voronoi cells with 5 UAVs.

for any $i = 1, \ldots, n$. In other words, at each step t of the algorithm, each UAV is moving toward the gravity center $Z_i[t, P_1, \ldots, P_n]$ as close as possible while satisfying the maximum displacement constraint (4.36).

Our analysis of the proposed navigation algorithm (4.43) requires some assumptions.

Assumption 4.4 Each UAV has the visibility radius R defined by (4.15). Moreover, it is assumed that at any time t, any point with the curvilinear coordinated z of the road C with $\rho_t(z) > 0$ is seen by at least one UAV. Also, each UAV knows the coordinates of other neighboring UAVs, so it can build its Voronoi cell.

Assumption 4.5 The curvature $k(z)$ of the curve C satisfies $|k(z)| < \frac{1}{R}$.

We model the density function $\rho_t(z)$ as the sum $\rho_t(z) = \alpha_t(z) + \beta_t(z)$ where $\alpha_t(z) = \rho_{t-1}(z - v_t)$ and $\beta_t(z)$ satisfies

$$\beta_t(z) = 0 \quad \forall z > z_I(t);$$
$$\beta_t(z) = 0 \quad \forall z < z_F(t);$$
$$\beta_t(z) < \rho_t(z) \quad \forall z_I(t) \leq z \leq z_F(t); \tag{4.44}$$
$$\int_{z_F(t)}^{z_I(t)} \beta_t(z)dz = 0.$$

Here v_t models the average displacement of the group of targets between time instants $t - 1$ and t, and $\beta_t(z)$ models the displacement of targets relative to each other. In particular, if $\beta_t(z) \equiv 0$ then all the targets make the same movement between the time instants $t - 1$ and t, and $\rho_{t-1}(z - v_t)$ is $\rho_{t-1}(z)$ shifted along C. Also, it is obvious that (4.44) implies that (4.34) is satisfied.

Assumption 4.6 There exist constants V_{max} and a such that

$$\|v_t\| \leq V_{max} \quad \forall t = 1, 2, \ldots$$

and

$$\int_{Z_1}^{Z_2} \beta_t(z)dz \leq a \quad \forall t = 1, 2, \ldots \quad \forall z_I(t) \leq Z_1 < Z_2 \leq z_F(t).$$

Moreover, $(1 + a)V_{max} < U_{max}$.

Now we are in a position to formulate the local optimality property of the proposed navigation algorithm.

Proposition 4.2 *Suppose that Assumptions 4.4, 4.5, and 4.6 hold. Then the trajectories $P_1(t), \ldots, P_n(t)$ of the UAVs (4.35) generated by the navigation algorithm (4.43) are locally optimal.*

Proof of Proposition 4.2: To deliver the minimum in (4.39) at any time t, all the UAVs should be located in R-neighborhood of the segment of the road containing the moving targets. Hence, the function $q(t, P_1(t), \ldots, P_n(t))$ is a continuous function on a compact set, which implies that the global minimum in the optimization problem (4.39) is achieved at some set of locations $(P_1^0(t), \ldots, P_n^0(t))$ [35]. Furthermore, it follows from Assumptions 4.4 and 4.5 that at any time t, the segment of the road containing moving targets consists of n subintervals $I_1(t), \ldots, I_n(t)$ such that for all the points of $I_i(t)$ the UAV i is the closest one at time t. Then it is obvious that

$$q(t, P_1(t), \ldots, P_n(t)) = \sum_{i=1}^{n} \int_{z \in I_i(t)} |(x_z, y_z), P_i(t)|^2 \rho_t(z) dz. \tag{4.45}$$

Furthermore, the minimum in

$$\min_X \int_{z \in I_i(t)} |(x_z, y_z), X|^2 \rho_t(z) dz$$

is achieved when the point X is at the gravity center of the corresponding Voronoi cell that is defined by (4.42). This and (4.45) together with Assumption 4.6 imply that the trajectories $P_1(t), \ldots, P_n(t)$ of the UAVs (4.35) generated by the navigation algorithm (4.43) are locally optimal. This completes the proof of Proposition 4.2. □

Remark 4.5 To implement the navigation algorithm (4.43), each UAV needs to compute the corresponding gravity center point $Z_i(t, P_1, \ldots, P_n)$. To do this, each UAV needs to know the density function $\rho_t(z)$ in its Voronoi cell $V_i(t, P_1, \ldots, P_n)$. Considering the last condition of (4.46), minor participation of a central controller is required to normalize the measured density functions. Specifically, each UAV sends its measured density function to the central controller, and the central controller returns the normalized density function to each UAV. The central controller can be one of the UAVs, and all the other UAVs keep reporting the measured density function periodically to this UAV.

4.6.3 Simulation Results

To demonstrate the convergence of the proposed method, we first consider a static case, where the targets are static and we can easily verify whether the UAVs move to the optimal positions. We set $U_{max} = 3$ m/s, $\phi_0 = \frac{\pi}{2}$, $h = 20$ m. Then, $R = 20$ m. The density function of a point on the curve C is represented by color in Figure 4.26a. The lighter the color, the larger the density. The dark parts, which are outside the interval $[z_I(t), z_F(t)]$, are with 0 density, see the left and right parts of the curve in Figure 4.26a. In the simulation result shown in Figure 4.26a, four UAVs start from some initial positions (marked by +), and they move according

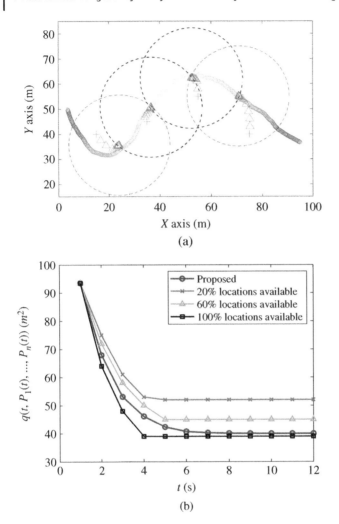

Figure 4.26 (a) The movements of 4 UAVs (marked by triangles). The initial positions are marked by +. The dash circles are the visible area at the final positions. A lighter position on the road presents a larger density. A video recording how these UAVs move is available here: https://youtu.be/ChKJJ31o8H4. (b) The objective function $q(t, P_1(t), \ldots, P_n(t))$ of the proposed method and the method of [46] using different percentages of locations. (b) Source: Adapted from [46].

to Eq. (4.54). It only takes 12 steps to converge (the termination condition is $||U_i||$ is smaller than a given threshold $\epsilon = 0.01$ for all $i = 1, \ldots, n$), and the objective function values are shown in Figure 4.26b.

We further consider a case where targets are mobile. We set $h = 13$ m and then $R = 13$ m. On the road shown in Figure 4.27a, we randomly deploy 150 vehicles. As in practice, the precise location of each vehicle is difficult to be obtained, we assume that each vehicle position follows a normal distribution. Thus, the density of any point on the road is the summation of the probability densities of the 150 vehicles. Note that different from the work in mobile sensor networks which has considered the measurement errors in the locations of sensors, the location information of UAVs is relatively high since they can have line-of-sight (LoS) to satellites with high probability. The measurement error in the considered system lies in the measurement of targets' locations, which are often estimated by image/video processing techniques. We consider that these vehicles are moving with a constant speed toward the right for 80 steps. Then, the density function also moves. The initial density function and the initial positions of four UAVs are shown in Figure 4.27a. As it is impossible to present the change of the density function over time, we provide a video in the link: https://youtu .be/oLuVUKzpSrM, which shows that the UAVs follow the vehicles. Figure 4.27b demonstrates the objective function values during the movement. Different from the static case where the UAVs converge to the optimal positions, in the mobile case, as the targets are moving, the optimal positions vary with time. As the UAVs can move faster than the targets, they can maintain the UAV-target distance and also the objective function to a low value.

Moreover, we regard the method of [46] as a benchmark method and apply it to the above-simulated scenarios. Though the method of [46] was originally designed for supporting wireless communications, it can still be applied here. The basic idea of [46] is to cluster the targets based on the precise locations and then assign UAVs to service the clusters. It is easy to understand that the precise location information of targets is difficult to be collected in practical applications. Here, we consider a certain percentage (20%, 60% and 100%) of targets' locations are available. For the situations of 20% and 60%, we randomly select the targets. When we evaluate the performance, all the targets are considered. In Figure 4.26b, we demonstrate the performance of the compared method on the static case. Clearly, when all the targets' locations are known, the optimization index $q(t, P_1(t), \ldots, P_n(t))$ achieves the lowest value. From Figure 4.26b we can see that the proposed method achieves very similar surveillance performance with the case of 100% targets' locations are given. The gap lies in that the proposed method makes use of the target density, which does not precisely capture the targets' locations. The proposed method significantly outperforms the compared method when only 20% and 60% targets' locations are known. For the case with mobile targets, the comparisons are shown in Figure 4.27b. Similar to Figure 4.26b, the compared method using all the targets' locations has the best surveillance performance, and the proposed method closely follows it. From these comparisons, we can conclude that the proposed method can achieve very good surveillance performance without using precise location information.

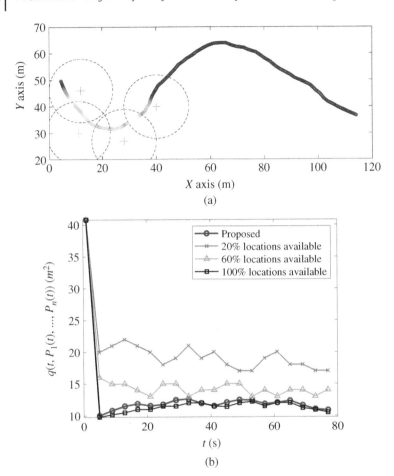

Figure 4.27 (a) Initial density function and UAV positions. A video recording how the targets and UAVs move is available here: source: Modified from https://youtu.be/oLuVUKzpSrM. (b) The objective function $q(t, P_1(t), \ldots, P_n(t))$ of the proposed method and the method of [46] using different percentages of locations. Source: Adapted from [46].

4.7 Navigation of UAVs for Surveillance of Groups of Moving Ground Targets

Following Section 4.6, this section also considers the case where targets can move. Different from Section 4.6, we consider that the targets can form groups. Different small groups can merge into a large group, and one group can also separate into some small groups.

4.7.1 Problem Statement and Proposed Approach

We consider that some groups of ground targets start from given initial positions, follow the predefined paths $\mathcal{P}$, merge at some position, and then move together, see Figure 4.28a. Besides the normal Cartesian coordinates, we introduce a z coordinate along the paths $\mathcal{P}$ corresponding to the length of a path segment. We can start from one point on the path and set its z coordinate as zero. All the points on the paths can have a unique z coordinate if we can construct a one-stroke graph for the paths by adding some extra street segments which are not parts of the paths. For example, in Figure 4.28b, we link the two initial locations with some extra street segments.

Suppose the system evolves at discrete times $t = 0, 1, 2, \ldots$. The moving targets are modeled by a density function $\rho_t(z)$ at time t. At t, $\rho_t(z)$ describes the positions of the targets. Since the considered moving targets only appear on a part of the paths, $\rho_t(z)$ satisfies the following conditions:

$$\rho_t(z) = 0 \quad \forall z \notin Z(t);$$

$$\rho_t(z) > 0 \quad \forall z \in Z(t);$$

$$\int_{z \in Z(t)} \rho_t(z)dz = 1,$$

(4.46)

where $Z(t) \in \mathcal{P}$ is the subset of the paths that has targets at time t, while $\rho_t(z)$ is zero when $z \notin Z(t)$.

The UAV team consists of $n \geq 2$ UAVs, and all of them are deployed at the same altitude $h > 0$. Let $P_1(t) = (x_1(t), y_1(t)), \ldots, P_n(t) = (x_n(t), y_n(t))$ denote the Cartesian coordinates of the UAVs at time $t = 0, 1, 2, \ldots$. The motion of the UAVs is described in Eq. (4.35). The displacement of a UAV follows Eq. (4.36).

Considering the fact that urban areas may have some buildings higher than h, the flight zone of UAVs is an area at the altitude of h excluding the tall buildings,[1] see Figure 4.28c. However, the zone excluding the tall buildings may not be always feasible for UAVs to see targets. When a tall building is between a UAV and a target, it cannot be seen even though it is inside the visibility circle. Thus, we introduce a connected, closed, and bounded deployment zone $\mathcal{D}$. This deployment zone satisfies the following property: for any points $d \in \mathcal{D}$ and $p \in \mathcal{P}$ such that $|d, p| \leq R$, all points of the straight line segment connecting d and p belong to $\mathcal{D}$. This property obviously guarantees that if a UAV is inside the deployment zone, it can see all the targets that are inside its visibility circle of radius R, so blockage of LoS is impossible. Notice that the deployment zone $\mathcal{D}$ may be non-convex.

---

1 A safety margin is used when we exclude the buildings. Thus, it is still safe when a UAV is on the boundary of the flight zone.

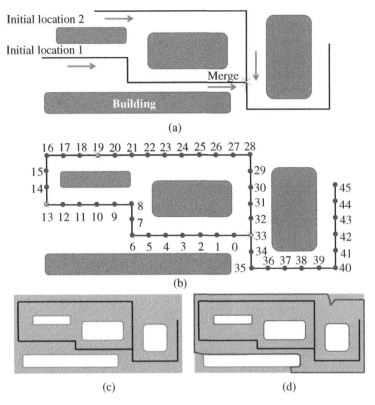

Figure 4.28 (a) Two groups of targets merge into one group. (b) z coordinate. (c) The flight zone (grey) that avoids collisions with buildings. (d) The deployment zone (the boundary is marked in dark grey) of UAVs that can see the path.

We assume that there exists a convex, closed, and bounded set $\mathcal{Z}$ such that $\mathcal{P} \subset \mathcal{Z}$, and the following constraint holds:

$$D := \max_{z \in \mathcal{Z}, d \in D} |z, d| > U_{max}. \tag{4.47}$$

Let (x_z, y_z) denote the Cartesian coordinates of a point on the path $\mathcal{P}$ with the coordinate z. The surveillance quality achieved by the UAVs placed at $P_1(t), P_2(t), \ldots, P_n(t)$ is quantified by the following function:

$$q(t, P_1(t), \ldots, P_n(t)) := \int_{z \in Z(t)} d^2_{min}(x_z, y_z) \rho_t(z) dz. \tag{4.48}$$

where $d^2_{min}(x_z, y_z)$ is defined by (4.37).

We consider the following optimization problem:

$$\min_{P_1(t), \ldots, P_n(t)} q(t, P_1(t), \ldots, P_n(t)) \tag{4.49}$$

where the minimum is taken over all possible UAV positions $P_1(t), \ldots, P_n(t)$ in $\mathcal{D}$. Note that minimizing (4.49) is equivalent to maximizing the surveillance quality

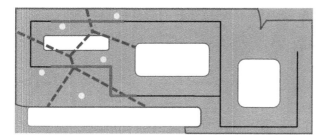

Figure 4.29 The Voronoi cells of five UAVs. The pink parts of the paths represent that there are targets there.

since a smaller value of (4.49) means that the moving targets are monitored from shorter distances, which is the same as (4.39). Moreover, all the UAVs must always be within D and any target's possible location must be seen by at least one UAV, i.e. $d_{\min}(x_z, y_z) \leq R, \forall z \in Z(t)$.

Let $Z_i(t, P_1, \ldots, P_n)$ be the subset of paths such that any point in this subset is closer to UAV i than other UAVs. Formally, when the UAVs are at $P_1, \ldots, P_n$, $Z_i(t, P_1, \ldots, P_n)$ is defined as the set of points $(x_z, y_z) \in P$ such that $|(x_z, y_z), P_i| \leq |(x_z, y_z), P_j|$ for all $j = 1, 2, \ldots, n$ and $j \neq i$, see Figure 4.29. In other words, $Z_i(t, P_1, \ldots, P_n)$ are so-called Voronoi cells. Let $O_i(t, P_1, \ldots, P_n) = (X_i(t, P_1, \ldots, P_n), Y_i(t, P_1, \ldots, P_n))$ be the gravity center of the cell $Z_i(t, P_1, \ldots, P_n)$, and

$$
\begin{aligned}
X_i(t, P_1, \ldots, P_n) &:= \frac{\int_{z \in Z_i(t, P_1, \ldots, P_n)} x_z \rho_t(z) dz}{\int_{z \in Z_i(t, P_1, \ldots, P_n)} \rho_t(z) dz}, \\
Y_i(t, P_1, \ldots, P_n) &:= \frac{\int_{z \in Z_i(t, P_1, \ldots, P_n)} y_z \rho_t(z) dz}{\int_{z \in Z_i(t, P_1, \ldots, P_n)} \rho_t(z) dz}.
\end{aligned}
\tag{4.50}
$$

Then the function (4.48) can be rewritten as follows[2]:

$$
\begin{aligned}
q(P_1, \ldots, P_n) &= \sum_{i=1}^{n} \int_{z \in Z_i} |(x_z, y_z), (x_i, y_i)|^2 \rho(z) dz \\
&= \sum_{i=1}^{n} \int_{z \in Z_i} ((x_z - x_i)^2 + (y_z - y_i)^2) \rho_t(z) dz \\
&= \sum_{i=1}^{n} \int_{z \in Z_i} (|(X_i, Y_i), (x_i, y_i)|^2 + (x_z^2 + y_z^2) \\
&\quad - (X_i^2 + Y_i^2) + 2x_i(x_z - X_i) + 2y_i(y_z - Y_i)) \rho(z) dz \\
&= \sum_{i=1}^{n} a_i |O_i, P_i|^2 + \sum_{i=1}^{n} b_i
\end{aligned}
\tag{4.51}
$$

---

2 Since (4.51) holds at any time, the time index t is omitted.

where $P_i = (x_i, y_i)$, $O_i = (X_i, Y_i)$ is the gravity center of Z_i given by (4.50),

$$a_i(t) = \int_{z \in Z_i} \rho(z) dz,$$

$$b_i(t) = \int_{z \in Z_i} (x_z^2 + y_z^2) \rho(z) dz - a_i(X_i^2 + Y_i^2). \qquad (4.52)$$

Note that $\sum_{i=1}^{n} \int_{z \in Z_i} 2x_i(x_z - X_i)\rho(z) dz = 0$ because $X_i = \frac{\int_{z \in Z_i} x_z \rho(z) dz}{\int_{z \in Z_i} \rho(z) dz}$; see (4.50). Similarly, $\sum_{i=1}^{n} \int_{z \in Z_i} 2y_i(y_z - Y_i)\rho(z) dz = 0$. It is clear from (4.51) that a desired situation is when P_i are close to the gravity centers $O_i(t, P_1, \ldots, P_n)$ of the corresponding Voronoi cells. In such situations, $\sum_{i=1}^{n} a_i|O_i, P_i|^2$ is small and the value of the cost function $q(P_1, \ldots, P_n)$ is close to $\sum_{i=1}^{n} b_i$. Therefore, we introduce the following definition.

Definition 4.6 Let $\epsilon > 0$ be a given constant. Trajectories $P_1(t), \ldots, P_n(t)$ of the UAVs are said to be locally ϵ-optimal if there exists a time $T > 0$ such that

$$q(t, P_1(t), \ldots, P_n(t)) < \sum_{i=1}^{n} b_i(t) + \epsilon \quad \forall t > T \qquad (4.53)$$

where $b_i(t)$ is defined by (4.52).

Note that like the standard definition of asymptotic stability of an equilibrium point of a nonlinear system from the theory of ordinary differential equations and modern control engineering, Definition 4.6 does not specify the rate of convergence to locally optimal values of the cost function. However, the simulations presented in this chapter will show that the proposed method leads to quite practical convergence times.

Problem Statement: For the given time-varying target density function $\rho_t(z)$, find the series of control inputs $u_1(t), \ldots, u_n(t)$ to construct locally ϵ-optimal trajectories $P_1(t), \ldots, P_n(t)$ of the UAVs in $\mathcal{D}$ satisfying constraint (4.36).

4.7.2 Navigation Method

The basic idea is partitioning paths with targets into subsets based on the current UAVs' locations, computing the gravity center of the density function in each subset, and then navigating the UAVs toward their corresponding gravity centers.

The gravity center $O_i(t, P_1, \ldots, P_n)$ may be outside $\mathcal{D}$, so let $\bar{O}_i(t, P_1, \ldots, P_n) \in \mathcal{D}$ be the nearest point to $O_i(t, P_1, \ldots, P_n)$ when $O_i(t, P_1, \ldots, P_n)$ is outside $\mathcal{D}$.

For any UAV i, the navigation algorithm is as follows:

$$u_i(t) = \begin{cases} O_i(t, P_1, \dots, P_n) - P_i(t), & \text{if (a) holds,} \\ \frac{U_{\max}(O_i(t,P_1,\dots,P_n)-P_i(t))}{|O_i(t,P_1,\dots,P_n),P_i(t)|}, & \text{if (b) holds,} \\ \bar{O}_i(t, P_1, \dots, P_n) - P_i(t), & \text{if (c) holds,} \\ \tilde{O}_i(t, P_1, \dots, P_n) - P_i(t), & \text{if (d) holds.} \end{cases} \tag{4.54}$$

We present more details for the four conditions in (4.54).

- Condition (a) is: $|O_i(t, P_1, \dots, P_n), P_i(t)| \leq U_{\max}$ and $O_i(t, P_1, \dots, P_n) \in D$. This means the gravity center $O_i(t, P_1, \dots, P_n)$ can be reached from $P_i(t)$ in one step.
- Condition (b) is: $|O_i(t, P_1, \dots, P_n), P_i(t)| > U_{\max}$ and $P_i(t) + \frac{U_{\max}(O_i(t,P_1,\dots,P_n)-P_i(t))}{|O_i(t,P_1,\dots,P_n),P_i(t)|}$ $\in D$. This means the gravity center cannot be reached in one step. Then, the UAV sets the maximum speed and moves along the direction defined by the unit vector $\frac{(O_i(t,P_1,\dots,P_n)-P_i(t))}{|O_i(t,P_1,\dots,P_n),P_i(t)|}$.
- Condition (c) is $O_i(t, P_1, \dots, P_n) \notin D$ and $|O_i(t, P_1, \dots, P_n), P_i(t))| \leq U_{\max}$, which means the gravity center is outside D and it is reachable in one step. In this case, the UAV is navigated to the closest to $O_i(t, P_1, \dots, P_n)$ point of the segment connecting $P_i(t)$ and $O_i(t, P_1, \dots, P_n)$, that is also in D, i.e. $\bar{O}_i(t, P_1, \dots, P_n)$, rather than $O_i(t, P_1, \dots, P_n)$).
- Condition (d) is $|O_i(t, P_1, \dots, P_n), P_i(t)| > U_{\max}$ and $P_i(t) + \frac{U_{\max}(O_i(t,P_1,\dots,P_n)-P_i(t))}{|O_i(t,P_1,\dots,P_n),P_i(t)|} \notin D$. In this case, the UAV moves at the point $\tilde{O}_i(t, P_1, \dots, P_n)$ of the same segment closest to $P_i(t) + \frac{U_{\max}(O_i(t,P_1,\dots,P_n)-P_i(t))}{|O_i(t,P_1,\dots,P_n),P_i(t)|}$ that is still in D.

All the UAVs follow the navigation algorithm (4.54) to decide their movement. Our analysis of this navigation algorithm requires the following assumptions.

Assumption 4.7 Initially, all the UAVs are at some positions inside the deployment zone D.

Assumption 4.8 At any time t, any point with the coordinated z of the path P with $\rho_t(z) > 0$ is seen by at least one UAV. Also, each UAV knows the coordinates of its neighbor UAVs, so it can build its Voronoi cell.

Assumption 4.9 There exists a time N such that at any time $t \geq N$, either condition (a) or (b) holds for any i.

Assumption 4.10 There exists a constant $\delta > 0$ such that $\delta < \frac{U_{\max}^4}{D^2}$ where D is defined by (4.47), and $\|q(t + 1, P_1, \dots, P_n) - q(t, P_1, \dots, P_n)\| < \frac{\delta}{2}$ and $\|\sum_{i=1}^n b_i(t + 1) - \sum_{i=1}^n b_i(t)\| < \frac{\delta}{2}$ for all t and all points $P_1, \dots, P_n \in D$.

Notice that the navigation law (4.54) and Assumption 4.8 guarantee avoiding LoS blockage. Indeed, according to (4.54), each UAV takes care of targets that are in its Voronoi cell. Since Assumption 4.8 guarantees that all the targets are seen by at least one UAV and this UAV is the closest to all the targets in its Voronoi cell, we have that the distances from the UAV to the targets in its Voronoi cell do not exceed R. Since the UAV is in the deployment zone D, this and the property of D imply that there is no blockage of LoS to all targets in the UAV's Voronoi cell.

Assumption 4.10 can be viewed as a constraint on the maximum targets' speed. Under this assumption and a scenario in which targets move in groups on some streets, at an initial time, the UAVs have all the targets in view if we use a sufficient number of UAVs. Moreover, following (4.54), the UAVs are navigated toward the gravity centers of the targets in their Voronoi cells. Then, it is intuitively clear, that in practice, the UAVs navigated by (4.54) usually do not lose targets. Therefore, Assumption 4.8 very often holds in practical situations. It should also be pointed out, that Assumptions 4.7–4.10 are technical assumptions required for a mathematically rigorous proof of the main theoretical result of the chapter. Computer simulations show that the proposed algorithm often works well even when these assumptions do not hold.

Now we are in a position to state our main result.

Proposition 4.3 *Suppose that Assumptions 4.7, 4.8, 4.9, and 4.10 hold. Let $\epsilon :=$ $4U_{\max}^2$. Then, trajectories $P_1(t), \ldots, P_n(t)$ of the UAVs generated by the navigation algorithm (4.54) are locally ϵ-optimal.*

Proof. Let N be an integer from Assumption 4.9. First, we prove the following

Claim: There exists a constant ϵ_0 such that for all $t > N$, if $q(t, P_1(t), \ldots, P_n(t)) > \sum_{i=1}^{n} b_i(t) + \frac{\epsilon}{2}$, then $q(t+1, P_1(t+1), \ldots, P_n(t+1)) - \sum_{i=1}^{n} b_i(t+1) < q(t, P_1(t), \ldots, P_n(t)) - \sum_{i=1}^{n} b_i(t) - \epsilon_0$.

Indeed, let $\mathcal{V}$ be some partitioning (not necessarily a Voronoi partitioning) that partitions $Z(t)$ into n subsets $V_1, \ldots, V_n$. We introduce the following function

$$Q(t, \mathcal{V}, P_1, \ldots, P_n) := \sum_{i=1}^{n} \int_{z \in V_i} |(x_z, y_z), P_i|^2 \rho_t(z) dz. \tag{4.55}$$

It is clear that $q(t, P_1, \ldots, P_n) \leq Q(t, \mathcal{V}, P_1, \ldots, P_n)$ for any points $P_1, \ldots, P_n$ and any partitioning $\mathcal{V}$, because in $Q(t, \mathcal{V}, P_1, \ldots, P_n)$ defined by (4.55), we take the squared distance from any point to some P_i, whereas in $q(t, P_1, \ldots, P_n)$ defined by (4.48), we take the squared distance from any point to the closest among $P_1, \ldots, P_n$. Therefore,

$$q(t, P_1(t+1), \ldots, P_n(t+1)) \leq Q(t, \mathcal{V}, P_1(t+1), \ldots, P_n(t+1)). \tag{4.56}$$

Moreover, it follows from (4.51) and (4.54) (as according to Assumption 4.9 either (a) or (b) holds) that

$$Q(t, \mathcal{V}, P_1(t+1), \ldots, P_n(t+1)) = \sum_{i \in \mathcal{I}(t)} a_i(|O_i(t), P_i(t)| - U_{\max})^2 + \sum_{i=1}^{n} b_i$$

$$= \sum_{i \in \mathcal{I}(t)} a_i(|O_i(t), P_i(t)|^2 - 2|O_i(t), P_i(t)|U_{\max} + U_{\max}^2) + \sum_{i=1}^{n} b_i$$

$$\leq \sum_{i \in \mathcal{I}(t)} a_i(|O_i(t), P_i(t)|^2 - U_{\max}^2) + \sum_{i=1}^{n} b_i \tag{4.57}$$

where $\mathcal{I}(t)$ is the set of i such that $|O_i(t), P_i(t)| > U_{\max}$. Notice that (4.57) is true because if either (a) or (b) holds, then it follows from (4.54) that $|O_i(t), P_i(t+1)| = 0$ if $i \notin \mathcal{I}(t)$. Since $a_i \geq 0$ and $\sum_{i=1}^{n} a_i = 1$, we have that $\sum_{i \notin \mathcal{I}(t)} a_i(t)|O_i(t), P_i(t)|^2 \leq U_{\max}^2 \sum_{i \notin \mathcal{I}(t)} a_i(t) \leq U_{\max}^2 = \frac{\epsilon}{4}$. From this and (4.51) we obtain that if $q(t, P_1(t), \ldots, P_n(t)) > \sum_{i=1}^{n} b_i(t) + \frac{\epsilon}{2}$ then

$$\sum_{i \in \mathcal{I}(t)} a_i(t)|O_i(t), P_i(t)|^2 > \frac{\epsilon}{4}. \tag{4.58}$$

Moreover, it is clear that $|O_i(t), P_i(t)| \leq D$ where D is defined by (4.47). This and (4.58) imply that

$$\sum_{i \in \mathcal{I}(t)} a_i(t) > \frac{\epsilon}{4D^2}. \tag{4.59}$$

Furthermore, it follows from (4.56), (4.57), and (4.58) that $q(t, P_1(t+1), \ldots, P_n(t+1)) - q(t, P_1(t), \ldots, P_n(t)) \leq -U_{\max}^2 \sum_{i \in \mathcal{I}(t)} a_i(t) < -\frac{U_{\max}^2 \epsilon}{4D^2}$. This, Assumption 4.10 and the equation $\epsilon = 4U_{\max}^2$ imply that $q(t+1, P_1(t+1), \ldots, P_n(t+1)) - \sum_{i=1}^{n} b_i(t+1) - q(t, P_1(t), \ldots, P_n(t)) + \sum_{i=1}^{n} b_i(t) < -\frac{U_{\max}^2 \epsilon}{4D^2} + \delta = -\frac{U_{\max}^4}{D^2} + \delta$. Now the statement of **Claim** with $\epsilon_0 := \frac{U_{\max}^4}{D^2} - \delta$ follows from this, since according to Assumption 4.10 $\frac{U_{\max}^4}{D^2} - \delta > 0$.

If $q(t, P_1(t), \ldots, P_n(t)) \leq \sum_{i=1}^{n} b_i(t) + \frac{\epsilon}{2}$, then it follows from (4.51), (4.54), (4.56), and (4.57) that $q(t, P_1(t+1), \ldots, P_n(t+1)) \leq q(t, P_1(t), \ldots, P_n(t)) \leq \sum_{i=1}^{n} b_i(t) + \frac{\epsilon}{2}$. This and Assumption 4.10 imply that $q(t+1, P_1(t+1), \ldots, P_n(t+1)) - \sum_{i=1}^{n} b_i(t+1) \leq \frac{\epsilon}{2} + \delta < \frac{\epsilon}{2} + \frac{U_{\max}^4}{D^2} < \frac{\epsilon}{2} + U_{\max}^2 = \frac{3\epsilon}{4} < \epsilon$ where the second last inequality holds due to (4.47). This and **Claim** imply local ϵ-optimality. Finally, it obviously follows from (4.54) and Assumption 4.7 that $P_i(t) \in \mathcal{D}$ for all i, t. $\qquad\square$

Remark 4.6 Notice that the proposed algorithm often does not lead to globally optimal trajectories as the cost function (4.48) is quite non-convex. Also, the system state is updated at discrete instants $t = 0, 1, 2, \ldots$. If we sample with some sampling period $\tau > 0$ so the system is updated at $t = 0, \tau, 2\tau \ldots$, the maximum UAV speed becomes $U_{\max} \tau$, hence, $\epsilon = 4U_{\max}^2 \tau^2 \to 0$ as $\tau \to 0$.

Remark 4.7 The proposed navigation method leads the UAVs toward denser positions having a higher density. The importance level of a position can be directly integrated as the weight of the position. For example, when someone is injured or some other special event occurs at some position, its importance level is increased. Then, one UAV may move toward the more important position to observe the event.

Remark 4.8 The proposed method has not considered the issue of information time-lags, since the wireless transmission between the UAVs can be much faster than the physical movements of targets and UAVs.

Remark 4.9 The proposed method assumes that all the targets are visible. When they spread across the entire path, this assumption may be violated. To locate the areas with more targets, the UAVs need to patrol the roads, exchange the sensed data, partition the targets, and then move toward the gravity centers of the corresponding partitions. Clearly, having all the targets in the view of UAVs at any time becomes impossible in such a case. One strategy is navigating the UAVs to patrol the path having targets to minimize the target visit frequency. Different from the conventional border patrol problem where the UAVs can fly at a constant speed, patrolling road segments having targets may require the UAVs to adjust their speeds since the number of targets changes when the targets and UAVs move. This is an interesting problem for future research.

4.7.3 Simulation Results

We first present simulation results of a case with four UAVs monitoring two groups of targets starting from different positions. One group (the above one in Figure 4.30a) has 400 targets, and the other has 200 targets. They start to merge after about 30 slots. We set $U_{max} = 25$ m/slot, $\phi_0 = \frac{\pi}{2}$, $h = 100$ m. Then, $R = 100$ m. Figure 4.30a–d shows the positions of UAVs in at some slots and a video demonstrating how the UAVs and targets move in the 89 slots is available at the caption of Figure 4.30. We consider a benchmark method for comparison. It deploys a number of static UAVs over the road segments where targets will appear. This is usually adopted in the traffic monitoring relevant publications mentioned previously. For the considered case, 9 UAVs are deployed at the same altitude $h = 100$ m, and they fully cover the road segments; see Figure 4.30a. We consider two metrics for evaluation: The coverage ratio and the average horizontal UAV-target distance are considered for evaluation. Figure 4.30e shows the coverage ratio at each time slot. While the proposed method achieves full coverage of all the targets, the static deployment loses some targets when they move. Figure 4.30f compares average distances. The proposed method allows

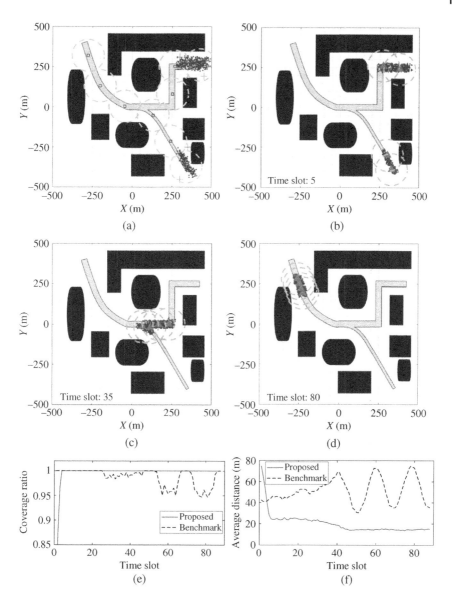

Figure 4.30 (a) Static positions of UAVs (black squares) and the visible ranges (dash circles). The initial positions of the UAVs by the proposed method (marked by +). (b–d) The movements of 4 UAVs (marked by triangles). The circles are the visible areas. A video recording how these UAVs move is available here: https://youtu.be/5y23Ox6wYjk. (e) Coverage ratio. (f) Average horizontal UAV-target distance.

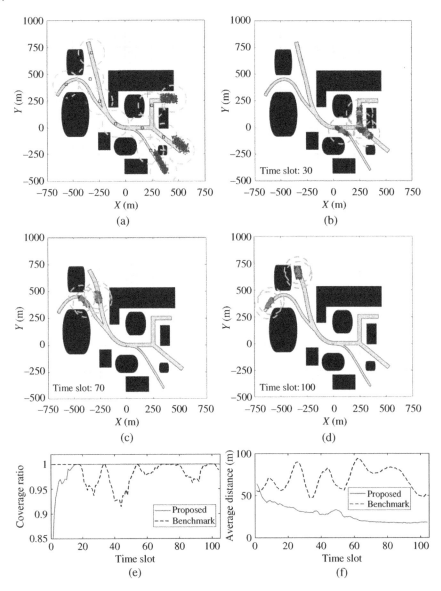

Figure 4.31 (a) The deployment of static UAVs. (b–d) The movements of 4 autonomous UAVs. A video showing how these UAVs move is available here: https://youtu.be/OEGr5-0zuyg. (e) Coverage ratio. (f) Average horizontal UAV-target distance.

UAVs to move toward a dense area for better surveillance. We understand that increasing the number of UAVs and reducing their separation in the static deployment can increase the coverage ratio and reduce the UAV-target distance, but the cost of surveillance is also increased.

We now consider a more complex case where three groups of targets merge and then separate; see Figure 4.31. We still use four UAVs. They fly a bit higher than the above case to view all the targets. Here, $h = 125$ m and then $R = 125$ m. We compare the static deployment with 11 UAVs. Figure 4.31b–d shows the positions of the UAVs at some slots. At slot 30, the groups start to merge, and at slot 70, they separate. How the UAVs move during the mission is shown in a video available in the caption of Figure 4.31. Similar to the last case, we consider the coverage ratio and the average horizontal UAV-target distance for evaluation. Figure 4.31e displays the coverage ratios achieved by the proposed method and the static deployment. The static deployment loses some targets when they are moving. At the worst time slot, about 8% of the targets cannot be viewed by the UAVs. If an event occurs at these lost targets, the system cannot detect it timely. Figure 4.31f shows the average UAV-target distance. Again, with the proposed algorithm, UAVs move closer to the targets.

4.8 Summary and Future Research

In this chapter, we discussed typical usages of UAVs in surveillance and monitoring. Firstly, we considered using a team of UAVs to monitor a moving disaster area. UAVs equipped with ground-facing cameras are to be deployed over the fastest-spreading parts of the frontier of the moving disaster area. A computationally efficient navigation algorithm was developed. Moreover, the proposed algorithm is optimal in the sense that the locations of the UAVs navigated according to this algorithm converge to the global maximum of the stated optimization problem. Secondly, UAVs are navigated over a ground region to periodically monitor every point of the region. An efficient navigation algorithm was developed. The developed algorithm is asymptotically optimal in the sense that the revisit period of the algorithm is getting close to the optimal revisit period as the size of the ground region tends to infinity. Thirdly, we presented a method to navigate a team of UAVs to monitor the traffic on a road segment. It consists of several stages: *initial, searching, accumulating*, and *monitoring*. The UAVs only need to share their positions and the measured information with their neighbors. Fourthly, we investigated the problem of navigation of a network of UAVs to maximise the quality of surveillance of a group of targets moving along a road with unknown time-varying speeds. A computationally simple and easily implementable real-time distributed navigation algorithm was proposed. Unlike many existing centralized navigation algorithms that require global information, in the proposed method, each UAV determines its motion based on some local information with minimal involvement of the central controller. Moreover, the convergence of the UAVs positions to the optimal locations was proved. Lastly, we addressed the problem of navigating a UAV team to maximize the surveillance quality for groups of moving ground targets that can merge into a large group or separate into more small groups. A computationally simple and easily implementable navigation algorithm was presented. It enables each UAV to determine its movement based on local information with minimal involvement of the central station, which outperforms many existing centralized algorithms that require global information. The local optimality of the proposed algorithm was proved. Computer simulations were conducted to evaluate the performance of the algorithms and comparisons with benchmark methods were presented. A limitation of the current work is that the proposed navigation laws are for a planar motion. Extending them to 3D scenarios is an important direction for our future research.

In this chapter and Chapter 3, the general aerial surveillance has been considered. In this type of applications, the UAVs stay above the targets of interest. In Chapter 5, we will consider an emerging scenario where the UAV needs to disguise his intention of surveillance from the targets.

References

1 P. Odonkor, Z. Ball, and S. Chowdhury, "Distributed operation of collaborating unmanned aerial vehicles for time-sensitive oil spill mapping," *Swarm and Evolutionary Computation*, vol. 46, pp. 52–68, 2019.

2 M. Messinger and M. Silman, "Unmanned aerial vehicles for the assessment and monitoring of environmental contamination: an example from coal ash spills," *Environmental Pollution*, vol. 218, pp. 889–894, 2016.

3 H. Cruz, M. Eckert, J. Meneses, and J.-F. Martínez, "Efficient forest fire detection index for application in unmanned aerial systems (UASs)," *Sensors*, vol. 16, no. 6, p. 893, 2016.

4 C. Yuan, Y. Zhang, and Z. Liu, "A survey on technologies for automatic forest fire monitoring, detection, and fighting using unmanned aerial vehicles and remote sensing techniques," *Canadian Journal of Forest Research*, vol. 45, no. 7, pp. 783–792, 2015.

5 L. Merino, F. Caballero, J. R. Martínez-De-Dios, I. Maza, and A. Ollero, "An unmanned aircraft system for automatic forest fire monitoring and measurement," *Journal of Intelligent and Robotic Systems*, vol. 65, no. 1–4, pp. 533–548, 2012.

6 H. X. Pham, H. M. La, D. Feil-Seifer, and M. C. Deans, "A distributed control framework of multiple unmanned aerial vehicles for dynamic wildfire tracking," *IEEE Transactions on Systems, Man, and Cybernetics: Systems*, vol. 50, no. 4, pp. 1537–1548, 2020.

7 H. Huang and A. V. Savkin, "Towards the internet of flying robots: a survey," *Sensors*, vol. 18, no. 11, p. 4038, 2018.

8 A. V. Savkin and H. Huang, "A method for optimized deployment of a network of surveillance aerial drones," *IEEE Systems Journal*, vol. 13, no. 4, pp. 4474–4477, 2019.

9 H. Huang and A. V. Savkin, "An algorithm of reactive collision free 3D deployment of networked unmanned aerial vehicles for surveillance and monitoring," *IEEE Transactions on Industrial Informatics*, vol. 16, no. 1, pp. 132–140, 2020.

10 A. V. Savkin and H. Huang, "Asymptotically optimal deployment of drones for surveillance and monitoring," *Sensors*, vol. 19, no. 9, p. 2068, 2019.

11 M. Papageorgiou, C. Diakaki, V. Dinopoulou, A. Kotsialos, and Y. Wang, "Review of road traffic control strategies," *Proceedings of the IEEE*, vol. 91, pp. 2043–2067, 2003.

12 M. Elloumi, R. Dhaou, B. Escrig, H. Idoudi, and L. A. Saidane, "Monitoring road traffic with a UAV-based system," in *IEEE Wireless Communications and Networking Conference (WCNC)*, pp. 1–6, 2018.

13 "These police drones are watching you," Accessed on 10 Nov. 2021. Online: https://www.pogo.org/analysis/2018/09/these-police-drones-are-watching-you/.

14 X. Liu, Z.-R. Peng, and L.-Y. Zhang, "Real-time UAV rerouting for traffic monitoring with decomposition based multi-objective optimization," *Journal of Intelligent and Robotic Systems*, vol. 94, no. 2, pp. 491–501, 2019.

15 F. Guerriero, R. Surace, V. Loscri, and E. Natalizio, "A multi-objective approach for unmanned aerial vehicle routing problem with soft time windows constraints," *Applied Mathematical Modelling*, vol. 38, no. 3, pp. 839–852, 2014.

16 H. Huang, A. V. Savkin, and C. Huang, "Decentralized autonomous navigation of a UAV network for road traffic monitoring," *IEEE Transactions on Aerospace and Electronic Systems*, vol. 57, no. 4, pp. 2558–2564, 2021.

17 A. V. Savkin and H. Huang, "Navigation of a network of aerial drones for monitoring a frontier of a moving environmental disaster area," *IEEE Systems Journal*, vol. 14, no. 4, pp. 4746–4749, 2020.

18 A. V. Savkin and H. Huang, "Asymptotically optimal path planning for ground surveillance by a team of UAVs," *IEEE Systems Journal*, pp. 1–4, 2021. https://ieeexplore.ieee.org/document/9580756.

19 A. V. Savkin and H. Huang, "Navigation of a UAV network for optimal surveillance of a group of ground targets moving along a road," *IEEE Transactions on Intelligent Transportation Systems*, pp. 1–5, 2021. https://ieeexplore.ieee.org/document/9430769.

20 H. Huang and A. V. Savkin, "Navigating UAVs for optimal monitoring of groups of moving pedestrians or vehicles," *IEEE Transactions on Vehicular Technology*, vol. 70, no. 4, pp. 3891–3896, 2021.

21 N. A. Mandellos, I. Keramitsoglou, and C. T. Kiranoudis, "A background subtraction algorithm for detecting and tracking vehicles," *Expert Systems with Applications*, vol. 38, no. 3, pp. 1619–1631, 2011.

22 L. Wang, F. Chen, and H. Yin, "Detecting and tracking vehicles in traffic by unmanned aerial vehicles," *Automation in Construction*, vol. 72, pp. 294–308, 2016.

23 S. Wang, F. Jiang, B. Zhang, R. Ma, and Q. Hao, "Development of UAV-based target tracking and recognition systems," *IEEE Transactions on Intelligent Transportation Systems*, vol. 21, no. 8, pp. 3409–3422, 2020.

24 A. Belkadi, H. Abaunza, L. Ciarletta, P. Castillo, and D. Theilliol, "Design and implementation of distributed path planning algorithm for a fleet of UAVs," *IEEE Transactions on Aerospace and Electronic Systems*, vol. 55, no. 6, pp. 2647–2657, 2019.

25 H. Oh, S. Kim, H. Shin, and A. Tsourdos, "Coordinated standoff tracking of moving target groups using multiple UAVs," *IEEE Transactions on Aerospace and Electronic Systems*, vol. 51, no. 2, pp. 1501–1514, 2015.

26 N. Karapetyan, K. Benson, C. McKinney, P. Taslakian, and I. Rekleitis, "Efficient multi-robot coverage of a known environment," in *2017 IEEE/RSJ International Conference on Intelligent Robots and Systems (IROS)*, pp. 1846–1852, IEEE, 2017.

27 Y. Gabriely and E. Rimon, "Spanning-tree based coverage of continuous areas by a mobile robot," *Annals of Mathematics and Artificial Intelligence*, vol. 31, no. 1, pp. 77–98, 2001.

28 H. Choset, "Coverage of known spaces: the boustrophedon cellular decomposition," *Autonomous Robots*, vol. 9, no. 3, pp. 247–253, 2000.

29 A. Yazici, G. Kirlik, O. Parlaktuna, and A. Sipahioglu, "A dynamic path planning approach for multirobot sensor-based coverage considering energy constraints," *IEEE Transactions on Cybernetics*, vol. 44, no. 3, pp. 305–314, 2014.

30 G. Leduc, "Road traffic data: collection methods and applications," *Working Papers on Energy, Transport and Climate Change*, vol. 1, no. 55, pp. 1–56, 2008. https://www.researchgate.net/publication/254424803_Road_Traffic_Data_Collection_Methods_and_Applications.

31 C. Yang, E. Blasch, J. Patrick, and D. Qiu, "Ground target track bias estimation using opportunistic road information," in *Proceedings of the IEEE 2010 National Aerospace & Electronics Conference*, pp. 156–163, IEEE, 2010.

32 R. Ke, Z. Li, J. Tang, Z. Pan, and Y. Wang, "Real-time traffic flow parameter estimation from UAV video based on ensemble classifier and optical flow," *IEEE Transactions on Intelligent Transportation Systems*, vol. 20, no. 1, pp. 54–64, 2019.

33 L. Cheng, L. Zhong, S. Tian, and J. Xing, "Task assignment algorithm for road patrol by multiple UAVs with multiple bases and rechargeable endurance," *IEEE Access*, vol. 7, pp. 144381–144397, 2019.

34 M. Li, L. Zhen, S. Wang, W. Lv, and X. Qu, "Unmanned aerial vehicle scheduling problem for traffic monitoring," *Computers & Industrial Engineering*, vol. 122, pp. 15–23, 2018.

35 A. N. Kolmogorov and S. V. Fomin, *Introductory real analysis*. New York: Dover, 1975.

36 Y. Choi, Y. Choi, S. Briceno, and D. N. Mavris, "Energy-constrained multi-UAV coverage path planning for an aerial imagery mission using column generation," *Journal of Intelligent and Robotic Systems*, vol. 97, no. 1, pp. 125–139, 2020.

37 S. Ann, Y. Kim, and J. Ahn, "Area allocation algorithm for multiple UAVs area coverage based on clustering and graph method," *IFAC-PapersOnLine*, vol. 48, no. 9, pp. 204–209, 2015.

38 N. L. Carothers, *Real analysis*. Cambridge University Press, 2000.

39 S. Rathinam, R. Sengupta, and S. Darbha, "A resource allocation algorithm for multivehicle systems with nonholonomic constraints," *IEEE Transactions on Automation Science and Engineering*, vol. 4, no. 1, pp. 98–104, 2007.

40 D. Kingston, R. W. Beard, and R. S. Holt, "Decentralized perimeter surveillance using a team of UAVs," *IEEE Transactions on Robotics*, vol. 24, no. 6, pp. 1394–1404, 2008.

41 Y. Kang and J. K. Hedrick, "Linear tracking for a fixed-wing UAV using nonlinear model predictive control," *IEEE Transactions on Control Systems Technology*, vol. 17, no. 5, pp. 1202–1210, 2009.

42 V. I. Utkin, *Sliding modes in control and optimization*. Springer Science & Business Media, 2013.

43 H. Shakhatreh, A. H. Sawalmeh, A. Al-Fuqaha, Z. Dou, E. Almaita, I. Khalil, N. S. Othman, A. Khreishah, and M. Guizani, "Unmanned aerial vehicles (UAVs): a survey on civil applications and key research challenges," *IEEE Access*, vol. 7, pp. 48572–48634, 2019.

44 I. R. Petersen, V. A. Ugrinovskii, and A. V. Savkin, *Robust control design using H^{∞} methods*. London: Springer-Verlag, 2000.

45 K. Harikumar, J. Senthilnath, and S. Sundaram, "Multi-UAV Oxyrrhis marina-inspired search and dynamic formation control for forest firefighting," *IEEE Transactions on Automation Science and Engineering*, vol. 16, no. 2, pp. 863–873, 2019.

46 B. Galkin, J. Kibilda, and L. A. DaSilva, "Deployment of UAV-mounted access points according to spatial user locations in two-tier cellular networks," in *Wireless Days (WD)*, pp. 1–6, March 2016.

5

Autonomous UAV Navigation for Covert Video Surveillance

5.1   Introduction

In Chapters 3 and 4, we have discussed using unmanned aerial vehicles (UAVs) to conduct aerial surveillance. In these chapters, we focus on the case where the targets being monitored are not influenced by the operations of UAVs. In other words, no matter where the target is under surveillance or not, it behaves normally. While many applications fall into this category, such as traffic monitoring, there are some practical situations where the operations of UAVs may impact how the targets under surveillance behave. A typical example is the surveillance of the criminal suspect. In this case, the UAVs are used to record targets' misbehaviors as court evidence in legal processes, which helps law enforcement to improve countermeasures or quickly response to incidents [1, 2]. A prominent challenge in this application is that the target can become aware of surveillance when the UAV flies in the currently available flying modes such as "Follow Me" and "Orbit". Once the surveillance is noticed, the target may take actions leading to a potential failure of the surveillance missions. Thus, the UAV needs to operate in a way that can disguise visually its surveillance intention toward the target.

The aforementioned application scenario leads to an interesting research direction, which is called covert video surveillance. Generally, we have a target vehicle or a person that moves based on its objective such as going to a destination. Executing a covert surveillance task, a UAV needs to not only have a continuous view of the target but also disguise itself. The UAV needs to avoid being noticed visually by the target. The UAV flies along a carefully planned trajectory to maintain a good and uninterrupted view of the target. The UAV also needs to disguise itself. Despite some commercially available UAV products can autonomously follow a target and shoot video, they have not considered the covertness.

In this chapter, we present our preliminary results for covert video surveillance by a single UAV. In the first approach, we first present how to quantify the disguising performance. From common sense, the UAV must be neither too close

Autonomous Navigation and Deployment of UAVs for Communication, Surveillance and Delivery,
First Edition. Hailong Huang, Andrey V. Savkin, and Chao Huang.
© 2023 The Institute of Electrical and Electronics Engineers, Inc. Published 2023 by John Wiley & Sons, Inc.

to the target as would increase the likelihood of being detected, nor too far away due to its limited sensing range. Additionally, the UAV does not monitor the target from a certain relative angle and distance, as this may lead to a high likelihood of being noticed. With these considerations, we present a metric to characterize the disguising performance. Then, we present an online trajectory planning to minimize the trajectory length while maximizing the disguising metric, subject to the aeronautics of the UAV and the target's presence in the sight of the UAV. The proposed approach was extended to the problem of covert UAV-on-UAV video tracking and surveillance [3].

The second approach originates from biology and is inspired by the concept of motion camouflage. Motion camouflage is a stealth behavior first described in [4] by Srinivasan and Davey and observed in hoverflies [4], territorial disputes of dragon-flies [5], and hunting falcons [6]. Such stealth behavior is based on the principle where a predator conceals its motion with respect to a prey. This is achieved by moving in such a way as to make the motion of the predator's image indistinguishable from the motion of the image of a stationary object in the retina of the moving prey. Since with such a motion strategy the predator induces no optical flow on the target's visual system, the target is unable to discern that the predator is moving. There is also evidence that humans can be tricked in the same way [7], and the motion camouflage phenomenon appears to be the reason for some motorbike accidents [8]. Motion camouflage guidance of UAVs can be of interest for policing and military applications, but it could also be used to observe animals in the wild or to record sport events with small UAVs without perceptual interference [9]. Based on the concept of motion camouflage, we present a bearing-only-based guidance method, which guides the UAV to conduct the covert video surveillance. The main results of this chapter were originally published in [10, 11].

The remainder of this chapter is organized as follows: Section 5.2 discusses some closely related publications. Section 5.3 presents the first approach including a formal problem statement, details of the proposed method, and its evaluation. Section 5.4 describes the second biologically inspired approach. Finally, Section 5.5 summarizes this chapter and points out several promising directions for future research.

5.2 Related Work

The research focusing on the normal aerial surveillance by UAVs is rich, i.e. a UAV monitors a target whose movement is independent of the UAV. References [12, 13] develop vision-based control methods that enable UAVs to track a moving ground target using video streaming. The objective is to keep a close UAV-target distance such that the target state can be updated frequently or guide the UAV to follow the motion of the target. References [14, 15] consider the problem of tracking a mobile target in urban environments, where the line-of-sight (LoS) between the UAV and the target is often blocked by buildings. This chapter focuses on the design of the UAV trajectory such that the probability of viewing the target is maximized. This chapter [16] proposes a coarse-to-fine UAV target tracking strategy with deep reinforcement learning to tackle the frequently changed aspect ratio of a target. These references all focus on computer vision [17]. Some other publications, such as [18, 19], account for the energy consumption in trajectory planning to extend the UAV's lifetime. When multiple targets are monitored, the movements of UAVs are often required to maximize the number of observable targets at any moment [20, 21], and the duration of monitoring when energy limitation is accounted [22]. Some companies have already released commercial UAV products, which can autonomously follow and monitor a target in the "Follow Me" mode, such as DJI MAVIC-2 (https://www.dji.com/au/mavic-2). Some other publications investigate the scenario where a UAV attempts to monitor a target that is aware of the UAV and tries to avoid being monitored [23–25]. Thus, a trajectory planning problem arises in an adversarial setting and depends on the motions of both the UAV and the target.

In a different, yet related context, trajectory planning and tracking are important research topics. For trajectory planning, Ref. [26] derives two classes of flight trajectories for tracking ground objects. The first class consists of switching between clockwise and counterclockwise orbits so that the average moving direction of the UAV is the same as that of the ground target. This class is suitable for tracking fast targets. The second class consists of orbits and straight lines, which is suitable for tracking slow targets. Reference [27] proposes an acceleration-continuous path-constrained trajectory planning algorithm with a built-in trade-off mechanism between the cruise motion and time-optimal motion. Regarding trajectory tracking, the study [28] develops a control Lyapunov function approach to address the problem of constrained nonlinear trajectory tracking control for UAVs. The study [29] presents a nonuniform control vector parameterization approach with time grid refinement for a flight-level tracking. One challenge of these approaches is the complex computation of the control inputs to the UAVs. Thus, some of the existing methods may not be suitable for real-time planning and tracking. A simple solution is to construct a trajectory for UAVs, which only consists of arcs and straight lines [30].

Regarding the motion camouflage strategy, different publications have presented computer simulations of motion camouflage, and several control mechanisms have been proposed. A finite-horizon linear quadratic regulator is used in [31] to generate the motion camouflage trajectories. It is assumed that the target moves at a constant velocity, and there are no motion restrictions on the predator. Under these assumptions, the problem is formulated as a time-independent linear system, and the controller becomes a time-dependent function. The study [8] presents a similar solution to [31]. A nonlinear indirect optimal control is used to generate motion camouflage for a target trajectory known beforehand. Neural networks have been used to achieve simulated motion camouflage [32]. The paper [8] presents a method using the polynomial NARMAX models [33] to learn a controller from motion camouflage training data, which allows interpreting the controller. Then, the authors implement the controller on mobile robots. In addition to ground robots, the study [34] proposes a guidance law by which a UAV can pursue a moving target from a constant distance while concealing its motion. A closed-form solution for the trajectory of the UAV is derived and tested via simulations.

5.3 Optimization-Based Navigation

In this section, we present an optimization-based navigation method. The basic idea is to model the covertness mathematically and then formulate the trajectory planning problem of a UAV as an optimization problem. Section 5.3.1 presents the system model, and Section 5.3.2 formally states the problem. Section 5.3.3 presents the trajectory planning method.

5.3.1 System Model

The considered system involves a UAV equipped with a camera and a target moving on a road. Suppose that the UAV flies at a fixed altitude h. Let $p(t) = (x(t), y(t))$ denote the coordinates of the UAV on the horizontal plane at time $t \in [0, T]$, where $[0, T]$ is the period of time during which the UAV executes a video surveillance task. The following model is considered for the UAV [35–37]:

$$\dot{x}(t) = v(t) \cos \theta(t),$$
$$\dot{y}(t) = v(t) \sin \theta(t), \tag{5.1}$$
$$\dot{\theta}(t) = \omega(t),$$

where $\theta(t) \in [0, 2\pi)$ is the UAV heading measured in the counterclockwise direction from the x-axis, $v(t) \in (0, V_{max}]$ is the linear speed, $\omega(t) \in [-W_{max}, W_{max}]$ is the angular speed, and V_{max} and W_{max} are the maximum linear and angular speeds, respectively. They depend on the mobility of the UAV. When $\omega(t) \neq 0$, the UAV makes a turn, and the turning radius is $r(t) = \frac{v(t)}{\omega(t)}$. The kinematic model (5.1) has been widely used to describe the motion of UAVs, see, e.g. [35–37]. Navigation algorithms developed based on the model or its minor variations have been successfully implemented in real-life experiments with UAVs, see e.g. [36, 38–41].

A robust control framework could be utilized in coupling with the proposed algorithm to suppress potential model uncertainties and disturbance. At the first stage of the framework, the proposed navigation/path planning algorithm is implemented. At the second stage, a robust controller, e.g. robust sliding mode controllers [38, 42], can be designed to reduce the effect of uncertainties and disturbances and keep the actual trajectory close to the one designed by the navigation/path planning algorithm.

Let $q(t) = (x_q(t), y_q(t))$ denote the position of the target at time t. Since the target moves on a road, the movement of the target is predictable for a certain period of time $\mathcal{T}$ into the future. We first assume that the predictions are perfect. At any time t_0, $q(t)$ is predicted for the time $t \in (t_0, t_0 + \mathcal{T}]$.

We consider the task of covert video surveillance. The UAV should not be too far away from the target, or the resolution of the video taken could degrade. The UAV should not be too close to the target either, since being too close would increase the possibility of being noticed by the target. We consider that when the UAV flies in a feasible flight zone, it can have the target in view with acceptable resolution. This region is behind the target and characterized by three parameters: R_1, the distance that the UAV needs to keep to avoid being noticed; R_2, the maximum distance for the UAV to maintain an acceptable resolution for its observation on the target, where $R_1 < R_2$; and an angle μ, see Figure 5.1a. We explain how to choose μ below.

Given the predicted position of the target at a future instant $(t + \delta)$, i.e. $q(t + \delta)$ (where δ is a sampling interval), and all the candidates of the UAV position at instant t, i.e. $p(t)$, we specify the feasible flight region of the UAV for instant $(t + \delta)$, as follows. The feasible flight region collects the intersections between the annulus specified by two concentric circles centered at $q(t + \delta)$ with radii of R_1 and R_2, and the disc centered at each candidate of $p(t)$ with radius equal to the maximum distance that the UAV can travel within a sampling interval δ. For ease of description, we define the angle μ in such a way that each feasible

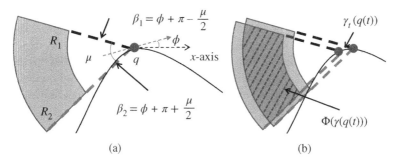

Figure 5.1 The feasible flight zone when the target movement prediction is (a) accurate, (b) inaccurate.

flight region at instant t is entirely captured in the part of the corresponding annulus specified by μ (referred to as the "feasible flight zone" for the instant t), and μ cannot be further reduced. At instant $(t + \delta)$, part of the feasible flight zone may not be reachable if the UAV is situated at a particular position at instant t. An additional constraint is in place to define an infinite cost for the UAV to fly from the particular candidate position at instant t to any candidate position in the unreachable part at instant $(t + \delta)$, as will be described in Section 5.3.3.

The feasible flight zone is a function of the target position and heading angle. Let $\phi(t)$ denote the moving direction of the target, which is measured from the x-axis in the counter-clockwise direction; see Figure 5.1a. This direction is tangent to the road at the position $q(t)$. Thus, given the road and the current target position, $\phi(t)$ can be readily computed. To further compute the feasible flight zone, we introduce two more angles β_1 and β_2, which are $\frac{\mu}{2}$ away from the opposite direction of the target moving direction ϕ: $\beta_1 = \phi + \pi - \frac{\mu}{2}$ and $\beta_2 = \phi + \pi + \frac{\mu}{2}$; see Figure 5.1a. Given q, μ, R_1, R_2, β_1, and β_2, the feasible flight zone can be obtained. Since μ, R_1, and R_2 are constants, β_1 and β_2 are two functions of ϕ, and ϕ depends on the target position $q(t)$, the feasible flight zone is a function of $q(t)$ given the road. Let $\Phi(q(t))$ denote the feasible flight zone when the target is at $q(t)$. The UAV must be within the feasible flight zone $\Phi(q(t))$:

$$p(t) \in \Phi(q(t)), \ \forall t \in [0, T]. \tag{5.2}$$

We assume that the onboard camera has a panoramic view or the capability of automatic tracking of targets. We also assume that a gimbal is used on the UAV to stabilize the camera.

When the predictions of the future target positions are not accurate, the inaccuracy can have impact on the feasible flight zone of the UAV. The feasible flight zone can become smaller when there is prediction uncertainty. Here, we assume that the prediction uncertainty is bounded per slot. This assumption is reasonable since the speed of the target is bounded in practice. Thus, instead of an estimated position, we have an estimated region, denoted by $\gamma(q(t))$, for a certain time instant t. For any point of an estimated region $q_0 \in \gamma(q(t))$ (see Figure 5.1b), we can compute a feasible flight zone that guarantees the view of q_0, i.e. $\Phi(q_0)$. Therefore, the feasible flight zone to view any point inside the region $\gamma(q(t))$ is the intersection of the feasible flight zones corresponding to all the interior points of $\gamma(q(t))$. Let $\Phi(\gamma(q(t)))$ denote this zone. The potential UAV position is limited to be within it; see Figure 5.1b:

$$p(t) \in \Phi(\gamma(q(t))) = \cap_{q_0 \in \gamma(q(t))} \Phi(q_0), \ \forall t \in [0, T]. \tag{5.3}$$

Remark 5.1 Given that the target vehicle moves continuously with a smooth trajectory, the feasible flight zone of the UAV changes continuously even when the target turns sharply (including a U-turn). On the other hand, the construction of feasible flight zones does have to take more careful consideration of the surrounding environment when the target is turning. For example, the feasible flight zones may have to extend vertically and reduce horizontally, so that the UAV can still keep its distance to the target while avoiding crashing into roadside buildings.

5.3.2 Problem Statement

An important issue to be addressed first is how to quantify the disguising performance of the UAV carrying out covert video surveillance. To avoid being noticed or detected by the target, a possible way for the UAV to disguise its intention is to change the UAV-target angle and distance as frequently and drastically as possible. Given the altitude of the UAV, we only consider the angle $\alpha(t)$ and the distance $d(t)$ between the UAV and the target on the horizontal plane. The UAV-target distance is the Euclidean distance between the UAV projection and the target on the horizontal plane; see Figure 5.2:

$$d(t) := \sqrt{(x(t) - x_q(t))^2 + (y(t) - y_q(t))^2}. \tag{5.4}$$

The UAV-target angle is defined as the angle between the direction of the target-UAV connection and the heading angle of the target ($\phi(t)$) on the horizontal plane; see Figure 5.2:

$$\alpha(t) := \arctan\left(\frac{y(t) - y_q(t)}{x(t) - x_q(t)}\right) - \phi(t), \tag{5.5}$$

where the x-axis can be selected such that $x(t) \neq x_q(t)$. The considered model can be extended to the case where the target moves along a road on an uneven terrain. The effect of the changing altitude of the target can be compensated by adjusting the altitude of the UAV. Such functionality is available in many commercial UAV products, such as DJI MAVIC-2 (https://www.dji.com/au/mavic2).

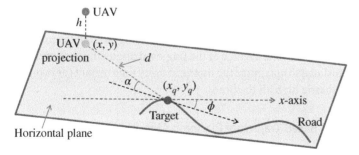

Figure 5.2 The UAV-target angle and distance. As all the notations in this figure are with the time index t, t is not shown.

Let $\eta > 0$ denote a given coefficient. The proposed disguising metric is given by a combination of the amplitudes of the derivatives of the UAV-target angle and the UAV-target distance:

$$\kappa(t) := \eta|\dot{\alpha}(t)| + |\dot{d}(t)|. \tag{5.6}$$

The rationale of the (5.6) is that, if the target carries cameras to monitor its surroundings, the UAV's frequently changing distance and angle with respect to the target also make it difficult for the target to correctly focus its camera lens. As a result, the target can hardly obtain a clear view of the UAV. When the driver or passenger of a target vehicle tries to visually detect possible trackers, e.g. the UAV, by naked eyes, flying in the defined feasible flight zone, which keeps a reasonable distance between the UAV and target, also makes it difficult for the target to see the UAV or differentiate it from the background.

The aim of disguising can be modeled as maximizing the disguising metric (5.6) over the time period of $[0, T]$:

$$\max_{p(t)} \int_{t=0}^{T} \kappa(t)dt. \tag{5.7}$$

Since the UAV is usually powered by a battery with limited capacity, high-energy consumption would result in a short lifetime. It is practically important to prevent the UAV from consuming too much energy. We use the length of the UAV trajectory to indicate the energy consumption. Then, the other goal can be formulated as follows:

$$\min_{p(t)} \int_{t=0}^{T} v(t)dt. \tag{5.8}$$

Given the locations of the target and parameters $R_1, R_2, \mu, \eta, T, W_{max}, V_{max}, p(0)$, and $\theta(0)$, the problem of interest is formulated as follows:

$$\min_{p(t)} \int_{t=0}^{T} v(t)dt - \lambda \int_{t=0}^{T} \left(\eta|\dot{\alpha}(t)| + |\dot{d}(t)| \right) dt, \tag{5.9}$$

s.t.

$$p(t) \in \Phi(\gamma(q(t))), \ \forall t \in [0, T], \tag{5.10}$$

where λ in (5.9) is a given weighting factor. It is worth pointing out that the integration in (5.9) is to evaluate the time average of the proposed disguising performance metric. It can be translated to maximize the metric at every time instant (or in other words, keeping the metric high all the time).

5.3.3 Predictive DP Based Trajectory Planning Algorithm

The problem under consideration is difficult to be solved due to the second term of (5.9). This is because the operator $|\cdot|$ makes (5.9) nondifferentiable at point zero. Besides, $|\dot{\alpha}(t)|$ and $|\dot{d}(t)|$ are nonconvex with respect to $p(t) = (x(t), y(t))$. As the objective function (5.9) is a continuous function of $v(t)$, $\alpha(t)$, and $d(t)$, finding the analytical solution is not straightforward. Furthermore, it is difficult to accurately predict the movement of the target for the whole surveillance period at the beginning of the task. In this section, we propose a DP-based method to solve the problem and plan the trajectory of the UAV online. Such a method periodically constructs the local UAV trajectory based on the target movement prediction for a short period of future time $\mathcal{T}$.

We first discretize the system. Given the sampling interval δ, the period $[0, \mathcal{T}]$ can be discretized into a number of slots with equal length, and the movement of the target is predictable for $N\left(N = \lfloor \frac{\mathcal{T}}{\delta} \rfloor\right)$ slots. At any slot, the UAV can only construct a trajectory for the next N slots. All the notations defined in the continuous-time domain can be transferred to the discrete-time domain. At the kth slot $(k = 0, 1, \ldots, N)$, the position, heading angle, linear speed, and turning radius of the UAV are $p[k] = (x[k], y[k])$, $\theta[k]$, $v[k]$, and $r[k]$, respectively. The position and heading angle of the target are $q[k] = (x_q[k], y_q[k])$ and $\phi[k]$, respectively. The UAV-target angle and distance are $\alpha[k]$ and $d[k]$, respectively. The objective function (5.9) can be discretized as follows:

$$\min \sum_{k=0}^{N-1} (v[k]\delta - \lambda\eta|\alpha[k+1] - \alpha[k]| - \lambda|d[k+1] - d[k]|). \tag{5.11}$$

When the UAV needs to construct the next local trajectory, we regard current slot as the initial stage, i.e. stage 0. In this stage, the UAV has a known initial position $s = p[0]$. We look for N bounded control inputs $v[k]$ and $\omega[k]$, $k = 0, \ldots, N-1$ such that $p[k+1] \in \Phi(\gamma(q[k+1]))$ and (5.11) is minimized, where $p[k+1]$ is computed by a discretized form of (5.1):

$$\begin{aligned} x[k+1] &= x[k] + v[k]\cos\theta[k]\delta, \\ y[k+1] &= y[k] + v[k]\sin\theta[k]\delta, \\ \theta[k+1] &= \theta[k] + \omega[k]\delta. \end{aligned} \tag{5.12}$$

Clearly, given the state of the UAV at stage k, i.e. $x[k], y[k], \theta[k]$, we obtain the state in the next stage $k+1$ when applying the control inputs $v[k]$ and $\omega[k]$.

Let S_k denote the state space of stage k for the UAV. This state space is the feasible flight zone, i.e. $S_k = \Phi(\gamma(q[k]))$, see Figure 5.1b. Let $g(p[k], v[k], \omega[k])$ denote the state transition cost from state $p[k] \in S_k$ to state $p[k+1] \in S_{k+1}$ by applying the control inputs $v[k]$ and $\omega[k]$. The expression for $g(p[k], v[k], \omega[k])$ is given by

$$g(p[k], v[k], \omega[k]) = v[k]\delta - \lambda \; \eta|\alpha[k+1] - \alpha[k]| - \lambda|d[k+1] - d[k]|, \tag{5.13}$$

where $d[k+1]$ and $\alpha[k+1]$ can be computed based on $p[k+1]$ and $q[k+1]$ by replacing t with k in (5.4) and (5.5), respectively.

We grid each feasible flight zone, i.e. the state space, and the position of the UAV at any slot is on a grid point of the feasible flight zone. Thus, in each stage, the state space is finite. Let $p_i[k]$ denote the ith candidate for $p[k]$ in S_k, and $p_j[k + 1]$ denote the jth candidate for $p[k + 1]$ in S_{k+1}. $v_{ij}[k]$ and $\omega_{ij}[k]$ are the suitable values of the control inputs $v[k]$ and $\omega[k]$, which allow the UAV to fly from $p_i[k]$ to $p_j[k + 1]$. According to the definition of the function $g(\cdot, \cdot, \cdot)$, the state transition cost from $p_i[k]$ to $p_j[k + 1]$ is given by $g(p_i[k], v_{ij}[k], \omega_{ij}[k])$. For illustration convenience, we define $c_{ij}^k = g(p_i[k], v_{ij}[k], \omega_{ij}[k])$. In the rest of this section, we present the DP framework, and below, we detail how to compute $v_{ij}[k]$ and $\omega_{ij}[k]$ for the transition from $p_i[k]$ to $p_j[k + 1]$.

As mentioned in Section 5.3.1, not all states in a stage can transit to any state in the next stage. Thus, if we cannot find feasible values for $v[k]$ and $\omega[k]$, the corresponding state transition cost is set as infinity, i.e. $c_{ij}^k = \infty$. For the cost associated with the final stage N, we introduce a virtual final state f. The final cost of a state in the final stage N is $c_{if}^N = 0$, since we do not specify the final UAV position.

A local trajectory consists of $N + 1$ states. Besides the known initial state s, the rest of the N states are to be selected from the predicted N state spaces. Let $J_k(j)$ denote the minimum cost of a segment of the local trajectory from the initial state s to a state j in stage $k + 1$. Then, the DP algorithm takes the following form [43]:

$$J_0(j) = c_{sj}^0, \quad j \in S_1, \tag{5.14}$$

$$J_k(j) = \min_{i \in S_k} \{c_{ij}^k + J_{k-1}(i)\}, \quad j \in S_{k+1}, \forall k = 1, \dots, N - 1. \tag{5.15}$$

Once J_{N-1} has been addressed, the optimal cost is given by

$$J_N = \min_{i \in S_N} \left\{ c_{if}^N + J_{N-1}(i) \right\}. \tag{5.16}$$

This is a forward DP algorithm. In stage 0, it records the subtrajectory cost from the initial state s to a state in stage 1. From stage k, the algorithm uses the already obtained minimum subtrajectory cost J_{k-1} to construct a new subtrajectory by adding a single-state transition cost. This repeats until all the states in stage $N - 1$, i.e. $J_{N-1}(j), j \in S_N$, are evaluated. As the final cost is 0 for any state in the final stage, (5.16) can be simplified by $J_N = \min_{i \in S_N} J_{N-1}(i)$. Now, we can obtain the minimum cost of the optimal trajectory.

To construct the trajectory, we further need a standard backtracking algorithm [44]. Indeed, the predecessor position and the linear speed of each minimum subtrajectory cost $J_k(j)$ should be recorded accordingly. Given the initial heading angle and the positions and linear speeds found by the backtracking algorithm, the trajectory can be constructed for the UAV. An illustrative example is provided in Figure 5.3, where there are three stages, and each stage has two states. We calculate the minimum cost from the initial state s to each state in a stage. At each state in a stage, the state which is in the previous stage and transits to the

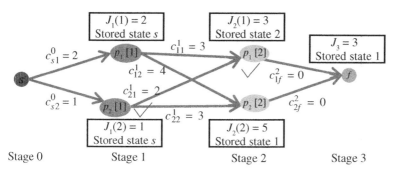

Figure 5.3 An illustrative example of constructing the path.

current state with the minimum cost is recorded. For example, state 1 of stage 1 records state s of stage 0, and state 1 of stage 2 records state 2 of stage 1. The path incurring the minimum end-to-end cost is backtracked by recursively selecting the preceding state associated with the minimum cost from the final stage, i.e. f, till the initial state s. In Figure 5.3, the waypoints of the identified path are state s of stage 0, state 2 of stage 1, and state 1 in stage 2.

Remark 5.2 In general, $N < \lfloor \frac{T}{\delta} \rfloor$, i.e. the UAV can only construct a trajectory based on the predictions of the target locations for the next upcoming N slots, instead of the whole trajectory for the period of $[0, T]$. When the predictions of the target position are accurate, the UAV constructs the trajectory for every N slots. However, when the measured target positions are outside the predicted region, e.g. when the target turns onto another road, the UAV reconstructs the trajectory using the updated current position and the predictions based on the current measurement.

5.3.3.1 Aeronautic Trajectory Refinement

In Section 5.3.3, we propose the DP framework to construct the UAV trajectory. In this section, we present a local trajectory design method with the consideration of the UAV kinematic model, inspired by the popular Dubins curves [45]. In particular, we opt to determine $v[k]$ and $\omega[k]$ such that the UAV can move from $p[k]$ with the heading angle $\theta[k]$ to $p[k + 1]$. With $v[k]$ and $\omega[k]$, $g(p[k], v[k], \omega[k])$ in (5.13) can be computed.

There are various ways to design $v[k]$ and $\omega[k]$. Our design is that at $p[k]$ the UAV starts to rotate at the beginning of the slot (if necessary) with either W_{max} or $-W_{max}$ (we explain how to select it after Remark 5.3) and a linear speed $v[k]$. When the UAV is at a position where its heading is toward $p[k + 1]$, the UAV sets the angular speed as zero [30]. Such a position is a key point of the trajectory, and as will be shown later, it is a tangent point. Then, it moves toward $p[k + 1]$ with the linear speed $v[k]$ along a straight line. Thus, the UAV applies a constant linear speed $v[k]$ for the whole slot and the maximum angular speed (either W_{max} or $-W_{max}$) for part of the slot; see Figure 5.4a. We will also explain how to determine the linear speed $v[k]$ in the final part of this section. This design simplifies not only the control of the UAV but also the problem to find $v[k]$ and $\omega[k]$, because it leads to a problem with only one variable, i.e. $v[k]$. Under such a design, the trajectory consists of an arc and a straight line segment; see Figure 5.4a. Note that if at $p[k]$, the UAV is heading toward $p[k+1]$, there is no arc on the trajectory. It is also worth pointing out that selecting the maximum angular speed (either W_{max} or $-W_{max}$) leads to the shortest possible trajectory. As shown in Figure 5.4b, given $p[k]$, $\theta[k]$ and $p[k+1]$, with the decreasing angular speed, the turning circle radius becomes larger and the trajectory becomes longer. For a UAV modeled by (5.1), applying either W_{max} or $-W_{max}$ results in the minimum turning circle and the shortest trajectory.

Remark 5.3 Despite its relevance to the popular Dubins car model [46], this design is different in the sense that the linear speed can be adjusted here, offering more flexibility than the Dubins car model with a fixed linear speed. The adjustable linear speed further leads to a variable turning radius, which also differs from the fixed radius in the Dubins car model.

Now, we consider how to determine the turning direction, i.e. the selection of W_{max} or $-W_{max}$. The UAV can either turn right or left at $p[k]$. Then, there are two feasible turning circles on both sides of the heading angle $\theta[k]$; see Figure 5.4a. As there are two tangent lines from an outside point to the circle, there exist four sets of trajectories to reach $p[k+1]$. Two of them are not feasible since the headings at the end of the arcs are not consistent with the corresponding tangent lines; see Figure 5.4a. Moreover, as one of our goals is to minimize the length of the UAV trajectory, the UAV should choose a direction such that the trajectory is shorter than that turning in the other direction. Such a decision is easy to make: if $p[k+1]$ is on the left of the heading angle $\theta[k]$, the UAV makes a left turn; otherwise, it makes a right turn. For example, in Figure 5.4a, $p[k+1]$ is on the right of the heading $\theta[k]$, thus the UAV should turn right and move toward $p[k+1]$.

To facilitate the description of this decision-making, we introduce some more symbols, and we demonstrate them in Figure 5.4. Let $\sigma[k] \in [0, 2\pi)$ denote the angle from the x-axis to the vector $\overrightarrow{p[k]p[k+1]}$ in the counter-clockwise direction; see Figure 5.4a. Then, the angle from the heading $\theta[k]$ to the direction of the vector $\overrightarrow{p[k]p[k+1]}$ in the counter-clockwise direction is given by $\sigma[k] - \theta[k] \in (-2\pi, 2\pi)$. The UAV determines the turning direction by the following rule:

$$\begin{cases} \text{no turn, if } \sigma[k] - \theta[k] = 0; \\ \text{turn right, if } \sigma[k] - \theta[k] \in [-\pi, 0) \cup [\pi, 2\pi); \\ \text{turn left, if } \sigma[k] - \theta[k] \in (-2\pi, -\pi) \cup (0, \pi). \end{cases} \quad (5.17)$$

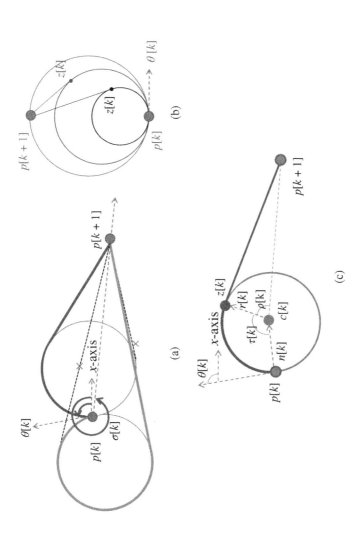

Figure 5.4 (a) UAV trajectories that can reach $p[k+1]$ from $p[k]$ with heading angle $\theta[k]$. (b) A larger angular speed results in a shorter trajectory. (c) Notations to build up the relations between $v[k]$ and the turning circle and the tangent line.

It is easy to show the correctness of (5.17). For a given heading $\theta[k]$, if $p[k+1]$ is on the left of $\theta[k]$, we can rotate $\theta[k]$ by an angle $\sigma[k] - \theta[k]$ in the counter-clockwise direction to get the direction $\sigma[k]$. Since we always wish to obtain a shorter trajectory, the rotated angle should be smaller than π, i.e. $\sigma[k] - \theta[k] \in (0, \pi)$. We can further subtract 2π on the right to obtain $\sigma[k] - \theta[k] \in (-2\pi, -\pi)$. This is the last rule in (5.17). The correctness of the second rule can be shown in the similar way, where we need rotate $\theta[k]$ in the clockwise direction. Note that when $\sigma[k] - \theta[k] = \pi$ or $-\pi$, turning left results in the same trajectory length with turning right.

Below, we present how to compute the linear speed $v[k]$ and the tangent point. $v[k]$ determines the turning radius given the angular speed, and the tangent point determines where the UAV stops turning. Let $n[k]$ be a unit vector that is perpendicular to the heading direction $\theta[k]$. $n[k]$ can be obtained by rotating $\theta[k]$ by $\frac{\pi}{2}$ either clockwisely or counterclockwisely (depending on the UAV turning direction). Let $c[k]$ be the center of the turning circle, and $z[k]$ denote the tangent point; see Figure 5.4c. We further denote $\tau[k]$ as the angle between the vectors $\overrightarrow{c[k]p[k]}$ and $\overrightarrow{c[k]z[k]}$; see Figure 5.4c. As the turning circle center $c[k]$ is $r[k]$ $\left(r[k] = \frac{v[k]}{W_{max}} \right)$ away from $p[k]$ along the direction $n[k]$, $c[k]$ can be computed by:

$$c[k] = p[k] + \frac{v[k]}{W_{max}} n[k]. \tag{5.18}$$

Since the tangent line $\overrightarrow{z[k]p[k+1]}$ is perpendicular to the vector $\overrightarrow{c[k]z[k]}$, we have:

$$\overrightarrow{c[k]z[k]} \cdot \overrightarrow{z[k]p[k+1]} = 0. \tag{5.19}$$

With $\overrightarrow{z[k]p[k+1]} = \overrightarrow{c[k]p[k+1]} - \overrightarrow{c[k]z[k]}$, $r[k] = |\overrightarrow{c[k]z[k]}|$, $r[k] = \frac{v[k]}{W_{max}}$ and (5.18), we can obtain the following equations:

$$\overrightarrow{c[k]z[k]} \cdot (\overrightarrow{c[k]p[k+1]} - \overrightarrow{c[k]z[k]}) = 0$$

$$\Rightarrow \overrightarrow{c[k]z[k]} \cdot \overrightarrow{c[k]p[k+1]} = |\overrightarrow{c[k]z[k]}|^2$$

$$\Rightarrow |\overrightarrow{c[k]z[k]}||\overrightarrow{c[k]p[k+1]}| \cos \rho[k] = |\overrightarrow{c[k]z[k]}|^2$$

$$\Rightarrow \cos \rho[k] = \frac{|\overrightarrow{c[k]z[k]}|}{|\overrightarrow{c[k]p[k+1]}|} = \frac{r[k]}{|\overrightarrow{c[k]p[k+1]}|} \tag{5.20}$$

$$\Rightarrow \rho[k] = \arccos \left(\frac{\frac{v[k]}{W_{max}}}{|p[k+1] - c[k]|} \right),$$

where $|\overrightarrow{\cdot}|$ is the length of a vector, and $\rho[k]$ is the angle between the vectors $\overrightarrow{c[k]z[k]}$ and $\overrightarrow{c[k]p[k+1]}$; see Figure 5.4c.

Furthermore, we can rotate the unit vector of $\overrightarrow{c[k]p[k+1]}$ by the angle of $\rho[k]$ in the counterclockwise direction to obtain the unit vector of $\overrightarrow{c[k]z[k]}$:

$$
\frac{\overrightarrow{c[k]z[k]}}{|\overrightarrow{c[k]z[k]}|} = \begin{bmatrix} \cos\rho[k] & -\sin\rho[k] \\ \sin\rho[k] & \cos\rho[k] \end{bmatrix} \frac{\overrightarrow{c[k]p[k+1]}}{|\overrightarrow{c[k]p[k+1]}|}. \tag{5.21}
$$

With $\overrightarrow{c[k]z[k]} = z[k] - c[k]$, $r[k] = |\overrightarrow{c[k]z[k]}|$, $r[k] = \frac{v[k]}{W_{max}}$ and (5.18), we obtain the expression for $z[k]$ as follows:

$$
z[k] = c[k] + \frac{v[k]}{W_{max}} \begin{bmatrix} \cos\rho[k] & -\sin\rho[k] \\ \sin\rho[k] & \cos\rho[k] \end{bmatrix} \times \frac{p[k+1] - c[k]}{|p[k+1] - c[k]|}. \tag{5.22}
$$

So far, we have derived the coordinates of the tangent point $z[k]$ as a function of $v[k]$ ($\rho[k]$ is already a function of $v[k]$ in (5.20)). To find $v[k]$, we consider the following equation, which requires the UAV to reach the point $p[k+1]$ by the end of the kth slot at a constant speed $v[k]$:

$$
|\overwidehat{p[k]z[k]}| + |\overrightarrow{z[k]p[k+1]}| = \delta v[k], \tag{5.23}
$$

where $|\overwidehat{p[k]z[k]}|$ is the length of the arc $\overwidehat{p[k]z[k]}$. Since $\tau[k]$ is the angle between the vectors $-n[k]$ and $\overrightarrow{c[k]z[k]}$, we have:

$$
\tau[k] = \arccos\left(\frac{-n[k] \cdot \overrightarrow{c[k]z[k]}}{|\overrightarrow{c[k]z[k]}|} \right). \tag{5.24}
$$

With $\tau[k]$ in (5.24) and $z[k]$ in (5.22), (5.23) can be re-written by

$$
\tau[k]\frac{v[k]}{W_{max}} + |p[k+1] - z[k]| = \delta v[k]. \tag{5.25}
$$

As all the other notations can be presented as functions of $v[k]$, (5.25) has only a single variable $v[k]$. By solving (5.25), we obtain the linear speed $v[k]$.

As mentioned, there are two tangent lines between a circle and an outside point. However, the above method cannot distinguish which one is feasible. Thus, we need to check once $z[k]$ is obtained by comparing the direction from $z[k]$ to $p[k+1]$ and the direction when the UAV leaves the circle at $z[k]$. If these two directions are the same, $z[k]$ is feasible. Otherwise, we need to replace $\rho[k]$ in (5.21) with $-\rho[k]$, which gives another tangent point. This tangent point is feasible for the UAV if the former is not.

For a brief summary, we have presented how to compute $v[k]$ given $p[k]$, $\theta[k]$ and $p[k+1]$. Once $v[k]$ is obtained, the tangent point $z[k]$ can be computed. The UAV follows the following procedures to move:

1. applies the maximum angular speed of either $-W_{max}$ or W_{max} decided by (5.17) at $p[k]$;
2. keeps turning until meeting $z[k]$;
3. sets the angular speed as zero and then moves in a straight line toward $p[k+1]$.

With $v[k]$, $g(p[k], v[k], \omega[k])$ in (5.13) can be calculated, so as the state transition cost. Then, the DP method presented in Section 5.3.3 can be realized to construct trajectory for the UAV. Since both the linear speed and the angular speed are within their respective bounds, the constructed trajectory is trackable by the UAV.

5.3.4 Evaluation

We study the performance of the proposed method via computer simulations using MATLAB. In this section, we first demonstrate the performance of the proposed method under different values of λ. Then, we investigate the impacts of several important factors, including target speed and road conditions, on the proposed method. The system parameters are set as follows: $\delta = 1$ second, $N = 5$, $\eta = 50$, $R_1 = 50$ m, $R_2 = 80$ m, $V_{max} = 20$ m/s, $h = 40$ m, $W_{max} = 1$ rad/s, $\mu = \frac{\pi}{4}$, $p[0] = [-60, 0]$ and $\theta[0] = -\pi$.

Following a given road, a target moves from left to right in a totally 90 slots (seconds), as shown in Figure 5.5. We demonstrate in Figure 5.5 some trajectories for different weighting factors λ ranging from 5 to 15. When $\lambda = 5$, the UAV moves

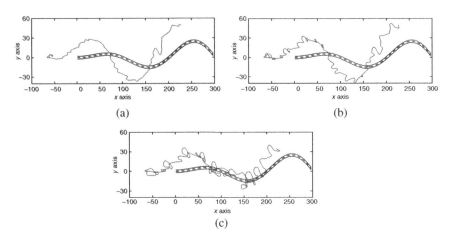

Figure 5.5 The UAV trajectories (solid line) to monitor a target (dash line) for different values of λ (results of one off simulations) (a) $\lambda = 5$. (b) $\lambda = 10$. (c) $\lambda = 15$.

Table 5.1 The average values of the linear speed $v(t)$, the derivative of the UAV-target angle $|\dot{\alpha}(t)|$, and the derivative of the UAV-target distance $|\dot{d}(t)|$, for the UAV movements under different values of λ.

λ	$v(t)$	$|\dot{\alpha}(t)|$	$|\dot{d}(t)|$
5	5.39	0.032	2.72
10	6.91	0.034	4.92
15	9.18	0.044	7.06

slightly in each slot; see Figure 5.5a; while when λ increases, the movement of the UAV becomes more significant; see Figure 5.5b, c. Table 5.1 summarizes the average values of the linear speed, the derivative of the UAV-target angle, and the derivative of the UAV-target distance in these movements, which are consistent with the trend shown in Figure 5.5. The UAV-target distance and angle are calculated based on (5.4) and (5.5), respectively. The derivatives of the UAV-target distance and angle are the differences between their respective values of two consecutive slots, given the sampling time $\delta = 1$. With the improvement of the disguising performance by increasing λ, the cost is the increased trajectory length (i.e. energy consumption). Since the on-board battery is limited in capacity, it requires a careful selection of λ to balance these factors.

We have conducted other simulations to derive some insights into the proposed method. Specifically, we have a look at how the movement of the target influences the movement of the UAV. Without loss of generality, we keep $\lambda = 10$. Similar results can be obtained for other values of λ. We first consider the speed of the target. On the same road as the above simulations, the target now finishes the movement in a totally 120 or 60 slots. The trajectories are shown in the first row of Figure 5.6. The trajectory when the UAV finishes the movement in 90 slots is shown here again for comparison. The average values of the disguising performance and the linear speed are shown in Figure 5.7a, b. From these trajectories and the average values of the considered metrics, we can see that with the increase of the target speed (i.e. decreasing the number of slots to finish the movement), the UAV's average linear speed increases, since it needs to guarantee the view of the target. Accordingly, the average derivatives of the UAV-target angle and distance increase as well. Thus, a fast speed of the target helps the UAV to disguise, because the UAV can adjust its trajectory easily to change the relative angle and distance between itself and the target. This is different from what we typically think, i.e. the faster the target is, the less likely it would be tracked.

Another interesting point influencing the method performance is road condition. In addition to the road simulated above, we create two other roads as shown in the second and third rows of Figure 5.6. They exhibit higher curvature than

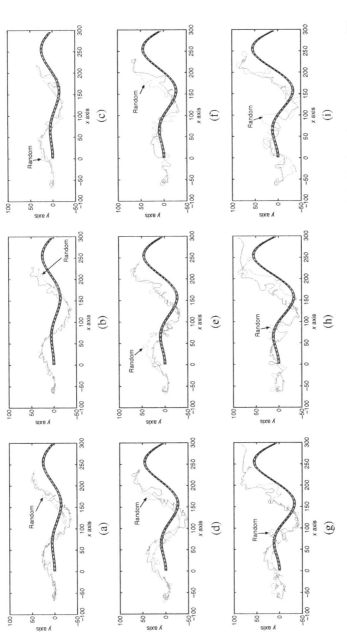

Figure 5.6 UAV trajectories by the proposed approach (solid) and a random benchmark method (dash) when the target moves on different roads in different speeds (results of one off simulations). (a) Road 1, 120 slots. (b) Road 1, 90 slots. (c) Road 1, 60 slots. (d) Road 2, 120 slots. (e) Road 2, 90 slots. (f) Road 2, 60 slots. (g) Road 3, 120 slots. (h) Road 3, 90 slots. (i) Road 3, 60 slots.

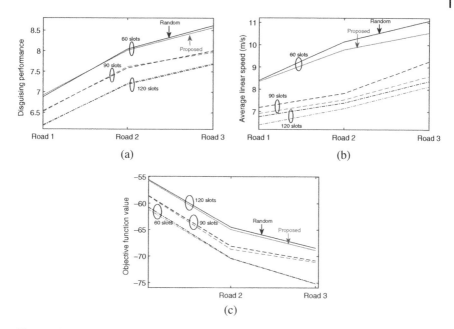

Figure 5.7 A comparison of the proposed method and the benchmark method (average results over 100 independent simulations) in terms of (a) the disguising performance, (b) the average linear speed, and (c) the overall objective function value for different roads and number of slots.

the first road. We simulate them under different numbers of slots to finish the movement. Correspondingly, the average values of the considered metrics are shown in Figure 5.7a, b. We see that under the same target speed and the weighting factor λ, the curvier the road is, the better disguising performance and the larger speed the proposed scheme provides. For the linear speed, as the target changes its heading angle more significantly on a curvier road, the feasible flight zone moves significantly as well. To make sure to have the target in its view, the UAV needs to fly faster. Also, the target heading (i.e. ϕ) relates to the UAV-target angle. Thus, a significant change of the target heading (i.e. a larger derivative of the target heading) can help increase the derivative of the UAV-target angle; see (5.5).

For summary, both a larger target speed and moving on a curvier road help the UAV disguise its intention. If the target moves slowly on a flat road, the weighting factor λ should be carefully selected by taking into account the expected disguising performance and the battery capacity.

To the best of our knowledge, the problem presented here is novel and has never been studied in the literature. To demonstrate the effectiveness of the proposed approach, we consider a random trajectory generation method as the benchmark for comparison purpose. Specifically, the benchmark method randomly selects a waypoint from each state space, then the proposed trajectory refinement method is applied to smoothly connect the waypoints. We conduct 100 independent simulations of the benchmark method under each of the configurations considered in Figure 5.6, and the average results of different metrics are presented in Figure 5.7. Figure 5.6 plots the representative trajectories produced by the proposed algorithm and the random benchmark under different configurations. In Figure 5.7, we see that the proposed method and the random method achieve comparable disguising performances. The advantage of the proposed method to the random method is that the average linear speed is lower in the proposed method than it is in the random method; see Figure 5.7a. As a result, the proposed method leads to smaller values of the objective function; see Figure 5.7b. As the parameter λ is selected to be relatively larger, the disguising performance is given priority in the trajectory generation process. To reflect realistic surveillance mission considerations, we have also conducted simulations in which the target moves on some other hand-crafted roads where the target makes large turns. The trajectories of the UAV and the results of linear speed and the derivatives of the UAV-target distance and angle are presented in Figures 5.8 and 5.9. Moreover,

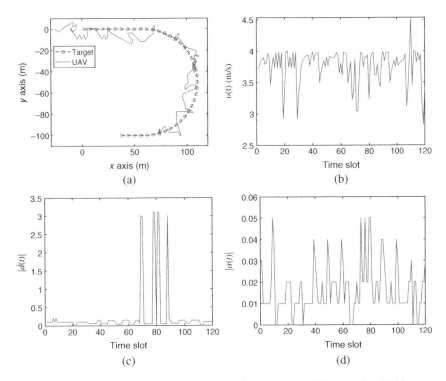

Figure 5.8 Simulation where the target moves along Road 4: (a) Trajectories; (b) Linear speed; (c) Derivative of relative distance; (d) Derivative of relative angle.

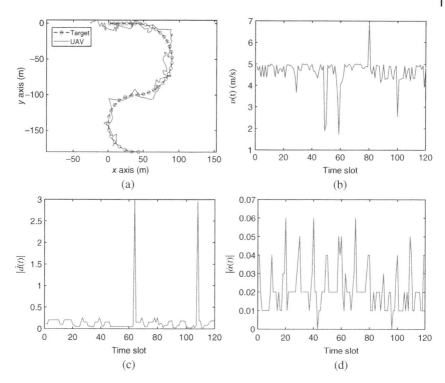

Figure 5.9 Simulation where the target moves along Road 5. (a) Trajectories. (b) Linear speed. (c) Derivative of relative distance. (d) Derivative of relative angle.

we conduct simulations where the target moves on real roads. In the case shown in Figure 5.10, the target does not follow a straight road. Instead, it moves onto another road at an intersection. We can see that the UAV's speed, angle, and distance change in the same way when the target is turning as they do when the target is moving along a straight line. As a result, the UAV can achieve consistent disguise performances throughout a surveillance mission, even when the target turns and changes roads.

Finally, the proposed algorithm is implemented and tested on a UAV simulator, CoppeliaSim. The UAV in the CoppeliaSim simulation is controlled by MATLAB via an already built API (see Figure 5.11). A target moves along a road, part of which is on a hill; see the thick solid curve Figure 5.12. The highest elevation of the road on the hill is 5 m. The UAV trajectory is indicated by the pink curve in Figure 5.12, where we consider two cases. In the first case, the altitude of the UAV remains unchanged with respect to the sea level; and in the second case, the altitude (i.e. the z-coordinate of the UAV) is adjusted so that the UAV can maintain its vertical distance to the target. The movements of the UAV and the target in

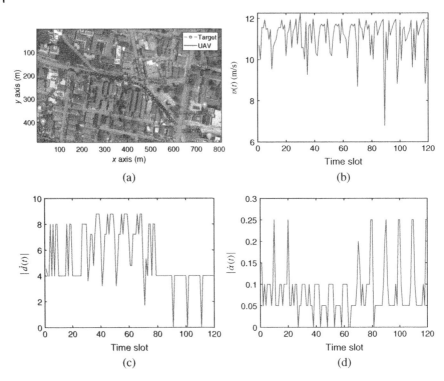

Figure 5.10 Simulation where the target moves along real roads: (a) Trajectories; (b) Linear speed; (c) Derivative of relative distance; (d) Derivative of relative angle.

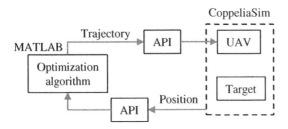

Figure 5.11 The interaction between the optimization algorithm in MATLAB and the simulation in CoppeliaSim.

the simulation are recorded in a video; see https://youtu.be/4zA8SPgE2Ec. We present the results of $|\dot{d}(t)|$ and $|\dot{\alpha}(t)|$ of these two trajectories in Figure 5.13. As seen in Figure 5.13a, the adjustment of the z-coordinate of the UAV adapting to the hilly terrain has little impact on the change of the relative UAV-target distance, i.e. $|\dot{d}(t)|$. The impact is nonnegligible on the change of the relative UAV-target angle $|\dot{\alpha}(t)|$, see Figure 5.13b. It is worth pointing out that in this simulation, the altitude

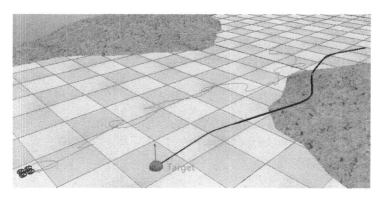

Figure 5.12 Simulation in CoppeliaSim. The movements of the UAV and the target in the simulation are recorded in https://youtu.be/4zA8SPgE2Ec. Source: Modified from https://youtu.be/4zA8SPgE2Ec.

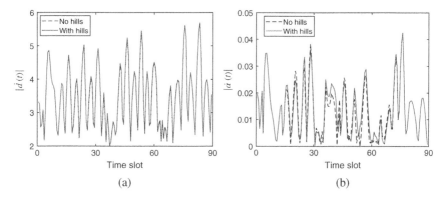

Figure 5.13 The absolute values of the derivative of UAV-target distance and UAV-target angle.

of the target on the path is relatively small. As the altitude of the target on the path rises, the adjustment of the UAV's z-coordinate (to maintain the vertical distance to the target) can be increasingly important to the disguising performance.

5.4 Biologically Inspired Motion Camouflage-based Navigation

In this section, we present another method for covert surveillance. This method is based on the well-known motion camouflage phenomenon. We state the problem in Section 5.4.1 and then present a sliding mode control method to achieve motion camouflage in Section 5.4.2.

5.4.1 Problem Statement

Different from Section 5.3, where the UAV flies at a fixed altitude (2D movement), we now consider 3D motion of a UAV described by the following mathematical model. Let

$$c(t) := [x(t), y(t), z(t)] \tag{5.26}$$

be the three-dimensional vector of the UAV's Cartesian coordinates. Then, the motion of the UAV is described by the equations:

$$\dot{c}(t) = v(t)a(t), \tag{5.27}$$

$$\dot{a}(t) = u(t), \tag{5.28}$$

where $a(t) \in \mathbf{R}^3$, $\|a(t)\| = 1$ for all t, $u(t) \in \mathbf{R}^3$, $v(t) \in \mathbf{R}$, and the following constraints hold:

$$\|u(t)\| \le U_{max}, \quad v(t) \in [-V_{max}, V_{max}], \tag{5.29}$$

$$(a(t), u(t)) = 0 \tag{5.30}$$

for all t. Here $\| \cdot \|$ denotes the standard Euclidean vector norm, $(\cdot, \cdot)$ denotes the scalar product of two vectors, the constants U_{max} and V_{max} are given. The scalar variable $v(t)$ and the vector variable $u(t)$ are control inputs, $v(t)$ is the speed or linear velocity of the 3D vehicle, $u(t)$ is applied to change the direction of the UAV's motion. The condition (5.30) guarantees that the vectors $a(t)$ and $u(t)$ are always orthogonal. Furthermore, $\dot{c}(t)$ is the velocity vector of the UAV. The kinematics of many UAVs can be described by the nonholonomic model (5.27), (5.28), (5.29), (5.30); see, e.g. [38] and references therein.

Our goal is to navigate the UAV to survey a moving ground target T. We assume that the target is moving on the ground that is not necessarily flat but might be quite uneven terrain. Therefore, the target's velocity $v_T(t)$ is a three-dimensional vector for any t such that $\|v_T\| \le V_T$, where $V_T > 0$ is some given constant. Furthermore, we select some fixed reference point P over the ground. In a practical situation, P may be some tall landmark such as the tip of a tower, etc.

5.4.1.1 Available Measurements

We assume that at any time t, the UAV has bearing measurements, i.e. directions from its current position to the moving ground target T and to the fixed reference point P, described by vectors $q(t)$ and $p(t)$ respectively; see Figure 5.14. We do not assume that the UAV knows the distance to the target T or the reference point P, as well as the target's velocity $v_T(t)$.

Definition 5.1 Let $D_2 > D_1 > 0$ be given constants. A guidance law is said to be a motion camouflage guidance law with parameters D_1, D_2 if there exists a time t_0

Figure 5.14 The measurements available at the UAV.

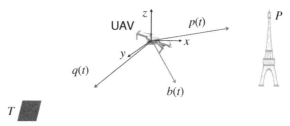

such that for any $t \geq t_0$, the UAV position at time t is on the straight line segment connecting the current target position and the point P. Moreover, the distance $d(t)$ between the UAV and the target T satisfies $D_1 \leq d(t) \leq D_2$.

Remark 5.4 Motion camouflage is a stealth behavior first described observed in many animals that allows a predator to successfully follow a prey while concealing its motion even without any advantage in speed [4, 6]. Since with such a motion strategy, the predator induces no optical flow on the prey's visual system, the target is unable to distinguish the moving predator from a steady object. Furthermore, the requirement of Definition 5.1 that the distance from the UAV to the target is bounded by some given constants, allows eliminating the effect of size with distance. In practice, D_1 and D_2 should be selected based on experimental results.

Our objective is to develop a motion camouflage, closed-loop guidance law based on bearing only measurements.

5.4.2 Motion Camouflage Guidance Law

Let w_1 and w_2 be nonzero three-dimensional vectors, and $\|w_1\| = 1$. Now, introduce the following function $F(\cdot, \cdot)$ mapping from $\mathbf{R}^3 \times \mathbf{R}^3$ to $\mathbf{R}^3$ as

$$F(w_1, w_2) := \begin{cases} 0, & f(w_1, w_2) = 0, \\ \|f(w_1, w_2)\|^{-1} f(w_1, w_2), & f(w_1, w_2) \neq 0. \end{cases}$$
$$f(w_1, w_2) := w_2 - (w_1, w_2)w_1. \tag{5.31}$$

In other words, the rule (5.31) defined in the plane of vectors w_1 and w_2, a vector $F(w_1, w_2)$ which is orthogonal to w_1 and directed "toward" w_2, see Figure 5.15.

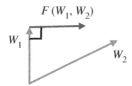

Figure 5.15 Illustration of the vector $F(w_1, w_2)$.

It is also clear that $F(w_1, w_2) = 0$ if w_1 and w_2 are colinear, and $\|F(w_1, w_2)\| = 1$ otherwise. Moreover, introduce the function $g(w_1, w_2)$ as follows:

$$g(w_1, w_2) := 1, \quad \text{if } (w_1, w_2) > 0,$$
$$g(w_1, w_2) := -1, \quad \text{if } (w_1, w_2) \le 0. \tag{5.32}$$

Introduce the vector $b(t)$ corresponding to the bisector of the angle between the vectors $q(t)$ and $p(t)$ that are the directions from the UAV to the reference point P and the target T, respectively; see Figure 5.14. We are now in a position to present the following sliding mode closed-loop guidance law:

$$u(t) = U_{max}g(a(t), b(t))F(a(t), b(t)),$$
$$v(t) = V_{max}g(a(t), b(t)). \tag{5.33}$$

In (5.33), control inputs are constantly updated based on bearing measurements to the moving target. Therefore, (5.33) is a closed-loop guidance algorithm.

Our main result requires the following assumptions:

Assumption 5.1 Let $L(t)$ be the distance between the reference point P and the target T at time t. Then $L_1 \le L(t) \le L_2$ for all t for some given constants $0 < L_1 < L_2$.

Let D_1 and D_2 be some constants such that $0 < D_1 < D_2 < L_1$.

Assumption 5.2 Let $\alpha := \min \left\{ \frac{L_1 - D_2}{D_2}, \frac{D_1}{L_2 - D_1} \right\}$. Then

$$\alpha V_{max} > \frac{D_1 V_T}{2L_1}, \tag{5.34}$$

$$U_{max} > \frac{V_T}{L_1}, \quad V_{max} > \frac{(L_2 - D_1)V_T}{L_1}, \tag{5.35}$$

$$D_1 + \frac{\pi(V_{max} + V_T)}{U_{max}} \le d(0) \le D_2 - \frac{\pi(V_{max} + V_T)}{U_{max}}. \tag{5.36}$$

Theorem 5.1 Suppose that Assumptions 5.1 and 5.2 hold. Then the UAV guidance law (5.33) is a motion camouflage guidance law with parameters D_1, D_2.

Proof. The guidance law (5.33) turns the velocity vector $\dot{c}(t)$ of the UAV toward the bisector $b(t)$ of the angle between the vectors $q(t)$ and $p(t)$ that are the directions from the UAV to the reference point P and the target T, respectively. Furthermore, it follows from (5.36) together with Assumption 5.1 that there exists some time $t_\star$ that for all $t \ge t_\star$ the vectors $\dot{c}(t)$ and $b(t)$ are colinear and $D_1 \le d(t) \le D_2$. Introduce the function $H(t)$ for all $t \ge t_\star$, where $H(t)$ is the distance between the UAV's current position and the straight line connecting P and the target's current position. Then, it follows from (5.34) of Assumption 5.2 and the inequality $D_1 \leq d(t) \leq D_2$ that if $H(t) > 0$, then $\dot{H}(t) < -\epsilon$ for some constant $\epsilon > 0$. Therefore, there exists a time t_0 such that the UAV's current position belongs to the straight line segment connecting P and the target's current position for all $t \geq t_0$. Moreover, Eq. (5.35) implies that for all $t \geq t_0$, the UAV will remain in the sliding mode of the system (5.27), (5.28), (5.33) corresponding to the position of the UAV on the straight line segment connecting P and the target's current position and the vector $\dot{c}(t)$ orthogonal to this straight line. This completes the proof of Theorem 5.1. □

Remark 5.5 Theorem 5.1 proves convergence of the UAV position to the straight line connecting the current target position and the point P. Moreover, Definition 5.1 and Theorem 5.1 also guarantee stability in the sense that the distance $d(t)$ from the UAV to the target satisfies $D_1 \leq d(t) \leq D_2$. It should be pointed out that the proof of Theorem 5.1 is analytical and mathematically rigorous. Theorem 5.1 also proves that the proposed closed-loop guidance algorithm is quite robust as it does not require any knowledge of the target's movement such as model, direction, and velocity. The only assumption is the upper limit on the target's speed. Moreover, we do not assume that the target moves on flat ground, and the ground may be uneven. It should be pointed out that Assumptions 5.1 and 5.2 are quite technical assumptions that are necessary for a mathematically rigorous proof of the performance of the proposed UAV guidance law. However, our simulations in Section 5.4.3 show that the guidance law (5.33) often performs well even in situations when they do not hold.

The proposed method cannot achieve the surveillance from a specific stand-off distance, since the range measurements and the target velocity knowledge are unavailable. Still, we can control the distance between the UAV and the reference P using the bearing measurement and the UAV's velocity.

5.4.3 Evaluation

We present some computer simulation results. We first consider that the target moves along an S-shape trajectory; see Figure 5.16, i.e. Case 1. The maximum speed of the target is $V_T = 1$ m/s. We set the maximum linear speed of the UAV as $V_{max} = 0.8$ m/s, $U_{max} = 0.5$, $D_1 = 15$ m, and $D_2 = 30$ m. The reference point $P = [-20, 0, 30]$. The initial position of the UAV is $[40, 20, 20]$. The trajectories of the UAV and the target during the simulation of 150 seconds is shown in Figure 5.16a, b. The initial position of the target is represented by the square, and the current positions of the target and the UAV are represented by the cube and the ball, respectively. We show the distance between the UAV and the target $d(t)$ in Figure 5.16d. From Figure 5.16d, we can see that after 10 seconds, the condition $D_1 \leq d(t) \leq D_2$ holds for any $t \geq t_0$. From Figure 5.16c, we can see that after $t_0 = 80$ seconds, the UAV is very close to the straight line segment connecting

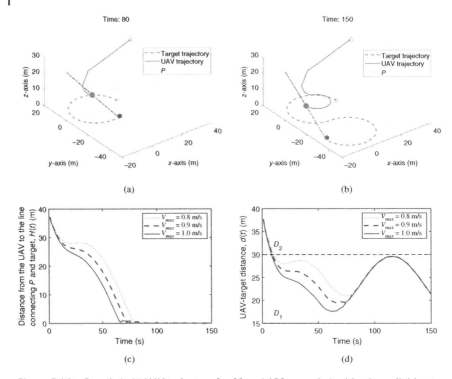

Figure 5.16 Case 1: (a, b) UAV trajectory for 80 and 150 seconds (a video is available at: https://youtu.be/mD0QsdNIml8). Source: Modified from https://youtu.be/mD0QsdNIml8. (c) The distance $H(t)$ from the UAV's current position to the line connecting P and the target. (d) The distance $d(t)$ between the UAV and the target.

the current target position and the reference point P. In other words, the proposed guidance law is a motion camouflage guidance law (see Definition 5.1). In our simulations, to avoid chattering (fast switching) that is typical for sliding mode controllers [47], we use the standard sliding mode control trick of approximating the sign type functions in (5.33) by a piecewise linear continuous saturation-type functions. Alternatively, throttle control can be considered to avoid chattering and reduce the control effort. We can also see that although $V_{max} < V_T$, the proposed method can still achieve the desired surveillance. We conduct more simulations with $V_{max} = 0.9$ m/s and $V_{max} = 1.0$ m/s. We show the distance from the UAV's current position to the line connecting P and the target $H(t)$ and the distance between the UAV and the target $d(t)$ in Figure 5.16c, d, respectively. We can see that larger V_{max} results in the earlier arrival on the line connecting P and the target. We demonstrate that the initial position of the UAV has little impact on the performance (see Case 2 in Figure 5.17).

Additionally, we consider the impact of the bearing measurement error of $q(t)$ on (i) $H(t)$, and (ii) the angle between the vectors $p(t)$ and $q(t)$. We add random noise

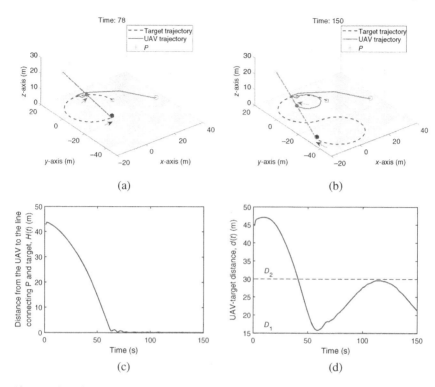

Figure 5.17 Case 2: UAV Trajectory for 78 and 150 seconds (a video is available at: https://youtu.be/9Sn9evhfMps). Source: Modified from https://youtu.be/9Sn9evhfMps. (a) The trajectories of the UAV and the target from 0 to 78 seconds. (b) The trajectories of the UAV and the target from 0 to 150 seconds. (c) The distance between the UAV and the line connecting P and the target during the simulation period. (d) The UAV-target distance during the simulation period.

to the measurement $q(t)$ and the amplitude of the noise is from 5% to 20% of the real measurement. For each value, we conduct 10 simulations independently. The results are shown in Figure 5.18a, b. Under the measurement noise, the average impact is small. For example, with 20% of measurement noise, the average distance gap of $H(t)$ is around 0.5 m, and the average angle gap is around 2°. The maximum distance gap is relatively large under 20% measurement noise, but the maximum angle gap, around 13°, is only 7% away from the true value, i.e. 180°. In other words, although the measurement has a large error, the impact on the disguising intention is relatively small.

Moreover, the proposed method is suitable for a target moving on an uneven terrain with a time-varying speed. Figure 5.19 reports this situation (Case 3). From Figure 5.19c, d, we can see that after $t_0 = 74$, the UAV is on the line segment connecting the target and P, and the condition $D_1 \leq d(t) \leq D_2$ holds, where D_1 takes 20 and D_2 takes 40.

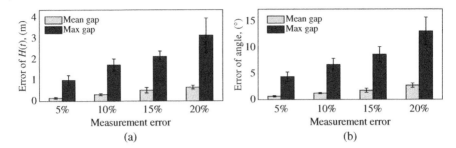

Figure 5.18 Impact of measurement errors.

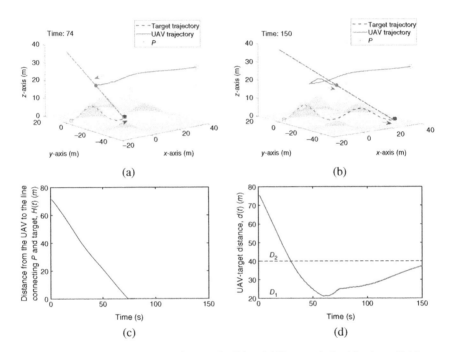

Figure 5.19 Case 3: (a-b) UAV trajectory for 74 and 150 seconds (a video is available at: https://youtu.be/htZVAJ9J8Pg). Source: Modified from https://youtu.be/htZVAJ9J8Pg. (c) The distance $H(t)$. (d) The distance $d(t)$.

5.5 Summary and Future Work

Beyond the normal aerial surveillance problems discussed in Chapters 3 and 4, in this chapter, we present recent approaches to covert video surveillance by a UAV. The first approach is based on a new metric characterizing the disguising performance, which evaluates the change of the relative distance and angle between the UAV and the target. An optimization is formulated to jointly maximize the disguising performance and minimize the energy efficiency of the UAV, subject to the motion constraint of the UAV and the requirement of keeping the target within view. A dynamic programming method is developed to plan the UAV's trajectory in an online manner. The second approach is based on the idea of motion camouflage, which is a stealth behavior. To achieve motion camouflage, the UAV keeps on a straight line segment connecting the target and a fixed reference point. A sliding mode control strategy is developed, which only takes the bearing information as input. So it is computationally light. Extensive computer simulations have been presented to demonstrate the effectiveness of these approaches.

There are several directions worth being considered further. First, since the approaches have only been evaluated through computer simulations, conducting field experiments with some volunteers would be interesting to test the effectiveness of the approaches in reality. Second, the methods developed in this chapter are only designed for a single UAV. Extending them to teams of UAVs is also interesting. Another important future research direction is an extension of the methods of this chapter to problems of covert collaborative eavesdropping on ground transmitters by teams of autonomous UAVs [48–51]. This chapter completes our discussion on the topic of aerial surveillance. In Chapter 6, we will focus on the usage of UAVs in parcel delivery.

References

1 A. Alipour-Fanid, M. Dabaghchian, N. Wang, P. Wang, L. Zhao, and K. Zeng, "Machine learning-based delay-aware UAV detection and operation mode identification over encrypted Wi-Fi traffic," *IEEE Transactions on Information Forensics and Security*, vol. 15, pp. 2346–2360, 2020.

2 "These police drones are watching you," Accessed on 20 Aug. 2019. Online: https://www.pogo.org/analysis/2018/09/these-police-drones-are-watching-you/.

3 S. Hu, W. Ni, X. Wang, A. Jamalipour, and D. Ta, "Joint optimization of trajectory, propulsion, and thrust powers for covert UAV-on-UAV video tracking and surveillance," *IEEE Transactions on Information Forensics and Security*, vol. 16, pp. 1959–1972, 2020.

4 M. V. Srinivasan and M. Davey, "Strategies for active camouflage of motion," *Proceedings of the Royal Society of London. Series B: Biological Sciences*, vol. 259, no. 1354, pp. 19–25, 1995.

5 A. Mizutani, J. S. Chahl, and M. V. Srinivasan, "Motion camouflage in dragonflies," *Nature*, vol. 423, no. 6940, p. 604, 2003.

6 S. A. Kane and M. Zamani, "Falcons pursue prey using visual motion cues: new perspectives from animal-borne cameras," *Journal of Experimental Biology*, vol. 217, no. 2, pp. 225–234, 2014.

7 A. J. Anderson and P. W. McOwan, "Humans deceived by predatory stealth strategy camouflaging motion," *Proceedings of the Royal Society of London. Series B: Biological Sciences*, vol. 270, no. 1, pp. S18–S20, 2003.

8 I. Ranó and R. Iglesias, "Application of systems identification to the implementation of motion camouflage in mobile robots," *Autonomous Robots*, vol. 40, no. 2, pp. 229–244, 2016.

9 I. Ra nó, "On motion camouflage as proportional navigation," *Biological Cybernetics*, vol. 116, no. 1, pp. 69–79, 2022.

10 H. Huang, A. V. Savkin, and W. Ni, "Online UAV trajectory planning for covert video surveillance of mobile targets," *IEEE Transactions on Automation Science and Engineering*, vol. 19, no. 2 pp. 735–746, 2022.

11 A. V. Savkin and H. Huang, "Bioinspired bearing only motion camouflage UAV guidance for covert video surveillance of a moving target," *IEEE Systems Journal*, vol. 15, no. 4, pp. 5379–5382, 2021.

12 S. Azrad, F. Kendoul, and K. Nonami, "Visual servoing of quadrotor micro-air vehicle using color-based tracking algorithm," *Journal of System Design and Dynamics*, vol. 4, no. 2, pp. 255–268, 2010.

13 F. Lin, X. Dong, B. M. Chen, K. Lum, and T. H. Lee, "A robust real-time embedded vision system on an unmanned rotorcraft for ground target following," *IEEE Transactions on Industrial Electronics*, vol. 59, pp. 1038–1049, 2012.

14 J. Kim and Y. Kim, "Moving ground target tracking in dense obstacle areas using UAVs," *IFAC Proceedings Volumes*, vol. 41, no. 2, pp. 8552–8557, 2008.

15 V. Shaferman and T. Shima, "Unmanned aerial vehicles cooperative tracking of moving ground target in urban environments," *Journal of Guidance, Control, and Dynamics*, vol. 31, no. 5, pp. 1360–1371, 2008.

16 W. Zhang, K. Song, X. Rong, and Y. Li, "Coarse-to-fine UAV target tracking with deep reinforcement learning," *IEEE Transactions on Automation Science and Engineering*, vol. 16, no. 4, pp. 1522–1530, 2019.

17 X. Huang, G. Mei, and J. Zhang, "Feature-metric registration: a fast semi-supervised approach for robust point cloud registration without correspondences," in *Proceedings of the IEEE/CVF Conference on Computer Vision and Pattern Recognition*, pp. 11366–11374, 2020.

18 S. C. Spangelo and E. G. Gilbert, "Power optimization of solar-powered aircraft with specified closed ground tracks," *Journal of Aircraft*, vol. 50, no. 1, pp. 232–238, 2012.

19 Y. Huang, H. Wang, and P. Yao, "Energy-optimal path planning for solar-powered UAV with tracking moving ground target," *Aerospace Science and Technology*, vol. 53, pp. 241–251, 2016.

20 H. Huang and A. V. Savkin, "An algorithm of reactive collision free 3-D deployment of networked unmanned aerial vehicles for surveillance and monitoring," *IEEE Transactions on Industrial Informatics*, vol. 16, pp. 132–140, 2020.

21 A. V. Savkin and H. Huang, "A method for optimized deployment of a network of surveillance aerial drones," *IEEE Systems Journal*, vol. 13, no. 4, pp. 4474–4477, 2019.

22 H. Huang and A. V. Savkin, "Reactive 3D deployment of a flying robotic network for surveillance of mobile targets," *Computer Networks*, vol. 161, pp. 172–182, 2019.

23 T. H. Chung, G. A. Hollinger, and V. Isler, "Search and pursuit-evasion in mobile robotics," *Autonomous Robots*, vol. 31, no. 4, p. 299, 2011.

24 S. A. Quintero and J. P. Hespanha, "Vision-based target tracking with a small UAV: optimization-based control strategies," *Control Engineering Practice*, vol. 32, pp. 28–42, 2014.

25 J. Ni and S. X. Yang, "Bioinspired neural network for real-time cooperative hunting by multirobots in unknown environments," *IEEE Transactions on Neural Networks*, vol. 22, no. 12, pp. 2062–2077, 2011.

26 R. W. Beard, "A class of flight trajectories for tracking ground targets with micro air vehicles," in *2007 Mediterranean Conference on Control & Automation*, pp. 1–6, IEEE, 2007.

27 P. Shen, X. Zhang, Y. Fang, and M. Yuan, "Real-time acceleration-continuous path-constrained trajectory planning with built-in tradeoff between cruise and time-optimal motions," *IEEE Transactions on Automation Science and Engineering*, pp. 1–14, 2020.

28 W. Ren and R. W. Beard, "Trajectory tracking for unmanned air vehicles with velocity and heading rate constraints," *IEEE Transactions on Control Systems Technology*, vol. 12, no. 5, pp. 706–716, 2004.

29 P. Liu, G. Li, X. Liu, L. Xiao, Y. Wang, C. Yang, and W. Gui, "A novel non-uniform control vector parameterization approach with time grid refinement for flight level tracking optimal control problems," *ISA Transactions*, vol. 73, pp. 66–78, 2018.

30 A. V. Savkin and H. Huang, "Optimal aircraft planar navigation in static threat environments," *IEEE Transactions on Aerospace and Electronic Systems*, vol. 53, pp. 2413–2426, 2017.

31 N. E. Carey, J. J. Ford, and J. S. Chahl, "Biologically inspired guidance for motion camouflage," in *2004 5th Asian Control Conference (IEEE Cat. No. 04EX904)*, vol. 3, pp. 1793–1799, IEEE, 2004.

32 A. J. Anderson and P. W. McOwan, "Model of a predatory stealth behaviour camouflaging motion," *Proceedings of the Royal Society of London. Series B: Biological Sciences*, vol. 270, no. 1514, pp. 489–495, 2003.

33 S. Chen and S. A. Billings, "Representations of non-linear systems: the NARMAX model," *International Journal of Control*, vol. 49, no. 3, pp. 1013–1032, 1989.

34 Y. Xu, "Analytical solutions to spacecraft formation-flying guidance using virtual motion camouflage," *Journal of Guidance, Control, and Dynamics*, vol. 33, no. 5, pp. 1376–1386, 2010.

35 S. Rathinam, R. Sengupta, and S. Darbha, "A resource allocation algorithm for multivehicle systems with nonholonomic constraints," *IEEE Transactions on Automation Science and Engineering*, vol. 4, no. 1, pp. 98–104, 2007.

36 D. Kingston, R. W. Beard, and R. S. Holt, "Decentralized perimeter surveillance using a team of UAVs," *IEEE Transactions on Robotics*, vol. 24, no. 6, pp. 1394–1404, 2008.

37 Y. Kang and J. K. Hedrick, "Linear tracking for a fixed-wing UAV using nonlinear model predictive control," *IEEE Transactions on Control Systems Technology*, vol. 17, no. 5, pp. 1202–1210, 2009.

38 C. Wang, A. V. Savkin, and M. Garratt, "A strategy for safe 3D navigation of non-holonomic robots among moving obstacles," *Robotica*, vol. 36, no. 2, pp. 275–297, 2018.

39 H. Li and A. V. Savkin, "Wireless sensor network based navigation of micro flying robots in the industrial internet of things," *IEEE Transactions on Industrial Informatics*, vol. 14, no. 8, pp. 3524–3533, 2018.

40 H. Li, A. V. Savkin, and B. Vucetic, "Autonomous area exploration and mapping in underground mine environments by unmanned aerial vehicles," *Robotica*, vol. 38, no. 3, pp. 442–456, 2020.

41 T. Elmokadem and A. V. Savkin, "A method for autonomous collision-free navigation of a quadrotor UAV in unknown tunnel-like environments," *Robotica*, vol. 40, no. 4 pp. 835–861, 2022.

42 A. S. Matveev, A. V. Savkin, M. Hoy, and C. Wang, *Safe robot navigation among moving and steady obstacles*. Elsevier, 2015.

43 R. Bellman, "Dynamic programming," *Science*, vol. 153, no. 3731, pp. 34–37, 1966.

44 R. E. Bellman and S. E. Dreyfus, *Applied dynamic programming*, vol. 2050. Princeton University Press, 2015.

45 H. Chitsaz and S. M. LaValle, "Time-optimal paths for a Dubins airplane," in *46th IEEE Conference on Decision and Control*, pp. 2379–2384, IEEE, 2007.

46 L. E. Dubins, "On curves of minimal length with a constraint on average curvature, and with prescribed initial and terminal positions and tangents," *American Journal of Mathematics*, vol. 79, no. 3, pp. 497–516, 1957.

47 V. Utkin, J. Guldner, and J. Shi, *Sliding mode control in electro-mechanical systems*. CRC Press, 2017.

48 H. Huang, A. V. Savkin, and W. Ni, "Navigation of a UAV team for collaborative eavesdropping on multiple ground transmitters," *IEEE Transactions on Vehicular Technology*, vol. 70, no. 10, pp. 10450–10460, 2021.

49 H. Huang and A. V. Savkin, "Energy-efficient decentralized navigation of a team of solar-powered UAVs for collaborative eavesdropping on a mobile ground target in urban environments," *AD Hoc Networks*, vol. 117, p. 102485, 2021.

50 H. Huang, A. V. Savkin, and W. Ni, "Decentralized navigation of a UAV team for collaborative covert eavesdropping on a group of mobile ground nodes," *IEEE Transactions on Automation Science and Engineering*, 2022. https://ieeexplore.ieee.org/document/9669934.

51 H. Huang, A. V. Savkin, and C. Huang, *Wireless communication networks supported by autonomous UAVs and mobile ground robots*. Elsevier, 2022.

6

Integration of UAVs and Public Transportation Vehicles for Parcel Delivery

6.1 Introduction

E-commerce has been playing a revolutionary role in the retail industry. In 2017, online sales were 2.3 trillion US dollars worldwide, which is 9.8% of the 23.45 trillion US dollars total global sales. It is projected to be 4.14 trillion US dollars in 2020, which may take 14.9% of the global sales [1]. As the suppliers need the means to transport the goods from their depots to their customers, such a huge market has made the logistics industry flourish. A direct result is the increasing use of ground vehicles in urban areas [2]. These vehicles not only congest the roads but also emit much pollutant gas. Thus, there is a clear need to design new methods of package distribution in urban areas. In recent years, unmanned aerial vehicles (UAVs) have been significantly advanced and found various civilian applications [3]. They have been recognized as a promising tool for the future logistics industry by many companies, such as Amazon [4], JD.com [5], and SF Express [6], as they are faster and more cost-efficient than the ground vehicles.

A general scenario of package distribution is that a supplier has a number of depots and each depot has a fleet of UAVs. Every a certain period (say a day), the supplier receives a list of orders. He needs to schedule the deliveries, i.e. deciding which depot to deal with which order, the departure time of a UAV, and what path a UAV should follow, so that some metric, such as the total cost, the makespan or the customer satisfaction level, is optimized. This mission consists of two tasks: high-level resource allocation, dealing with the order assignment, and low-level path planning for the UAVs, both of which depend on the delivery mode in use.

There are two popular UAV delivery modes. The first one exploits UAVs' mobility only. A UAV flies to a customer, drops off the parcel, returns to the depot, and recharges or replaces its battery. In this mode, the parcel-depot assignment and the parcel-UAV scheduling issues can be formulated as the conventional task-machine or job-shop scheduling problem, and can be addressed by many existing methods in operations research [7, 8]. One disadvantage of this mode is

Autonomous Navigation and Deployment of UAVs for Communication, Surveillance and Delivery,
First Edition. Hailong Huang, Andrey V. Savkin, and Chao Huang.

the UAVs' limited flight distance. As released by the Amazon Prime Air project, a typical Amazon UAV can fly about 32 km [4]. In other words, it covers roughly a circular area centered at the depot with a radius of 16 km. To fully cover a large area, a large number of depots may be necessary.

The second mode is called the UAV-vehicle collaboration [9–16]. A vehicle carries parcels and a UAV. The UAV can leave the vehicle, deliver a customer, and dock with the vehicle. In the meanwhile, the vehicle can deliver to other customers, and then move to the scheduled position to rendezvous with the UAV. The publications [9–16] are relevant to the conventional traveling salesman problem (TSP) and vehicle routing problem (VRP). Another type of UAV-vehicle collaboration uses UAVs to load ground vehicles [17]. The US UAV systems developer Matternet and the Swiss online marketplace siroop have developed a van and UAV-based system [18]. A UAV is loaded at the supplier and flies to the developed van. Then, the van driver takes possession of the parcel and delivers it to the customer, and the UAV returns to the supplier to pick up the next parcel. Different from the first mode, the second one can achieve the same delivery area as the conventional ground delivery. It is more energy-efficient than the pure ground delivery since UAVs reduce the usage of ground vehicles. However, its cost is high, since the driver needs to be paid, and the maintaining cost and the fuel cost of a ground vehicle are also higher than those of UAVs.

With these advantages and shortcomings, one may wonder if there exists a mode guaranteeing a large delivery area with a low running cost. There are several possible schemes. The first one is to optimally manage the battery usage [19] or develop a light power module with high capacity [20]. With the current battery technique, deploying charging stations is an option, where UAVs can recharge the battery and then fly again. Specifically, the references [21, 22] focus on the ground charging station deployment. This approach is relevant to the facility positioning problem. The third scheme, the focus of this chapter, exploits the transit network. A UAV is assumed to be able to ride on a public transportation vehicle on the roof and transfer between different vehicles; see Figure 6.1. It can reach a customer that locates far away from the depot, as long as some public transportation vehicle routes stretch there. The paper [23] concludes that the approach of exploring autonomous mobility, e.g. public transportation vehicles and even private cars, is feasible as an alternative delivery method in high-demand seasons.

The scheme of exploiting the transit network is more cost-efficient than the charging station construction method. Besides, the deliverable area can be significantly improved, since riding on a vehicle can significantly save energy for UAVs. However, it is also challenging since the UAVs need to cooperate with the vehicles. The public transportation vehicles have predefined routes and timetables, and they cannot wait for a UAV like a delivery vehicle [9]. Additionally, the travel times of public transportation vehicles are time-dependent. The travel

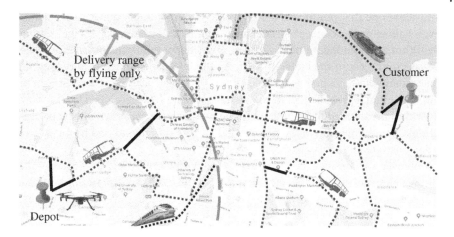

Figure 6.1 The illustration of the network combining the public transportation network and UAV flight network. The black solid segments represent the UAV flights. This approach significantly enlarges the UAV delivery range.

time of a vehicle between two stops in the peak hours may be longer than that in the off-peak hours. Moreover, the travel time is uncertain since no vehicles can exactly follow the given timetables. Therefore, to use these vehicles to transport UAVs, path planning algorithms that can deal with the time dependency and the stochastics are needed.

In this chapter, we discuss the recent development of the integration of UAVs and public transportation vehicles for parcel delivery. We first discuss the one-way path planning in a public transit network for UAV delivery. Following the network flow framework [24], the considered problem is formulated as an optimization problem to maximize the probability of delivering the customer on time. The so-called delivering on time requires the UAV to arrive at the customer within a given time. We characterize the path traversal time for different cases to extend a path. Specifically, we investigate the probability density functions (PDFs) of the UAV path when it is extended by adding a UAV flight and traveling with a vehicle. In the latter case, the UAV has to arrive at the stop node earlier than the vehicle with which the UAV plans to travel, and this involves a waiting time. Besides, with which vehicle a UAV can travel depends on the arrival instant of the UAV and the departure instant of the vehicle. Regarding this point, the work presented in this chapter differs from [25], which considers the reliable path planning in a bus network but assumes that the waiting time only depends on the travel time of the predecessor link. Attention is paid to the computation of the PDF of the UAV path in these cases. Following the concept of reliability in stochastic networks [26], the original probability maximization problem is converted to an equivalent one that minimizes the path traversal time under a given confidence level. Following the concept of stochastic dominance, we develop a label setting algorithm to find nondominant paths from the depot to the customer. Moreover, we take into account the battery lifetime of the UAV. The limited battery lifetime determines whether a path is feasible or not. Then, we have an extra constraint in both the original problem and the converted problem. The developed label setting algorithm is extended by taking one more operation to verify whether a path is feasible. The complexity of the proposed algorithm is analyzed and how it works is demonstrated by an extensive case study.

Moreover, we consider the round-trip routing problem for a UAV, which works for the deterministic case, i.e. the timetables of the public transportation vehicles are accurate. The main objective is to let the UAV finish the delivery in the shortest time so that it does not violate the battery constraint. To achieve this, a multimodal network consisting of public transportation vehicles' trips and UAV direct flights is first constructed. This multimodal network is hard to handle since an edge in the network may correspond to several vehicle services. We convert this complex network to a simple network in which any edge only represents one service. In this simple network, we formulate the shortest path problem to minimize the return time the UAV returns to the depot, subject to that the overall energy consumption on the path is below the initial energy. We present a Dijkstra-based method to find the shortest path in terms of time from the depot to the customer and then back to the depot. Since one of the most important advantages of the public transportation vehicle assisted UAV delivery approach is the increased coverage area, we present an analysis on the lower bound of the extended coverage.

We present another method for the round-trip routing problem for a UAV in the stochastic network. The problem is formulated to minimize the return instant to the depot subject to two practical constraints. The first one is that the arrival instant at the customer is before a certain deadline. Such a deadline may be required by the customer when he ordered, and the supplier also agreed with it. The second constraint is that the energy consumption of the UAV following this path is within a given budget to account for the limited battery capacity. To address this problem, an efficient algorithm is developed for the deterministic situation, and then it is extended to deal with the uncertain transit network. Its effectiveness is demonstrated via computer simulations. The main contribution is that the developed method can quickly compute the round-trip path for the UAV in the public transportation network while satisfying the delivery due time constraints and the energy budget constraint. This is one more step toward the large-scale application of using transit networks to assist UAV delivery. The main results of this chapter were originally published in [27–29].

The remainder of this chapter is organized as follows: Section 6.2 discusses some closely related references. Section 6.3 presents a general model used in this chapter. Section 6.4 develops an algorithm for one-way path planning. Section 6.5 proposes an algorithm for round-trip path planning in a deterministic network. Section 6.6 presents an algorithm for round-trip path planning in a stochastic network. Finally, Section 6.7 summarizes this chapter and points out several promising future research directions.

6.2 Related Work

For the last-mile parcel delivery, there exist three typical approaches. The traditional and common approach is by trucks. Starting from the depot, a truck visits a set of customers one by one and returns to the depot. To optimize the truck trajectory, TSP and its variants like asymmetric traveling salesman problem (ATSP) [30], clustered traveling salesman problem (CTSP) [31], TSP with time windows (TSP-TW) [32], and VRP [33] are often used.

Regarding UAV-vehicle collaboration, a typical method is to use a truck to carry several UAVs. This collaboration reduces the delivery time compared to the scheme using trucks only. The paper [9] is the earliest publication presenting this scheme, which formulates two problems: Flying Sidekick TSP and Parallel UAV Scheduling TSP. Some variants of [9] have also been published, such as TSP with UAV (TSP-D) [13], minimum cost TSP [15], TSP with UAV Station (TSP-DS) [14], heterogeneous delivery problem (HDP) [10], and VRP for UAV [11].

Parcels can be categorized into four groups in terms of size and weight and the distance between the customer and the depot, see Figure 6.2. The large and/or heavy parcels have to be delivered by trucks, and the small and light parcels that are near the depot can be delivered by UAVs. For those small and light parcels and the customers are far from the depot, trucks have to be used because they are outside the delivery area of UAVs. To avoid using trucks for these parcels, a possible way is to enlarge the delivery area. One method is to position charging stations [21, 22, 34–36], such that UAVs can charge their battery or replace the battery to prolong the flight. This requires construction costs and some facilities may have low utility. The paper [37] proposes the scheme of using a public transportation

Figure 6.2 The classification of parcels based on size, weight, and the distance between the customer and the depot, and what type of vehicle will be assigned.

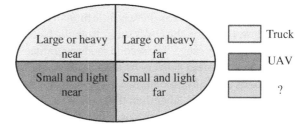

vehicle running on a fixed and closed trajectory such as a train to carry parcels and UAVs for parcel delivery. The UAVs can depart from the vehicle at some positions, fly to the customers and return to the vehicle at some other positions. This scheme ensures the delivery to the customers in a cost-efficient way, but the customers that can be served should be close to the trajectory.

Public transportation vehicles are recognized as a sustainable smart city enabler. They are natural mobile platforms traversing urban areas and enable many environmentally friendly applications. For example, a public transportation vehicle can take sensors to gather air pollutant levels [38]. A data sink installed on a vehicle can collect sensory data from sensors nodes deployed near its trajectory [39]. Beyond the internet-of-things applications, the paper [40] proposes to use public transportation vehicles to transport UAVs for the surveillance task. Inspired by this, the references [27, 41] propose the public transportation vehicle-based UAV delivery scheme, and the reference [23] even proposes to take the private vehicles into account for UAV delivery because private vehicles cover a larger area than the trajectory-fixed transportation vehicles. Since the public transportation vehicles and private cars (though they are with much more uncertainty) already exist on the roads, exploiting them in the results of the aforementioned applications at little extra cost. UAVs, also known as Unmanned Aerial Vehicles (UAVs), have advanced significantly and been applied widely in civilian domains such as wireless communication service [42], surveillance [43], parcel delivery [9], etc. In recent years, many logistics companies have started to test their UAV delivery systems [4, 6, 44, 45]. In the research community, two schemes relating to UAV delivery have been proposed. The first one focuses on only using UAVs for delivery: a depot is equipped with a fleet of UAVs, and each UAV generally follows the routine of recharging, flying to a customer, dropping off the parcel, and returning to the depot. The problem considered for this scheme is the scheduling problem [11, 46, 47]. The second one is called the UAV-truck collaboration scheme [9, 10, 14, 48]. Departing from the depot, a truck carries a set of parcels and one or several UAVs. The UAVs can launch from the truck, deliver customers, and then dock with the truck at some position after completing the deliveries. In the meanwhile, the truck can deliver to other customers, and then move to the scheduled position to rendezvous with the UAVs.

These two strategies both have merits and demerits. The first method can deliver a customer quickly, and as there is little human labor involved, the system running cost is low. However, the UAVs suffer from the battery constraint, which results in a limited delivery range. The second mode guarantees the same delivery area as the conventional ground delivery by trucks. However, it leads to a high running cost, since a driver is still needed. Besides, the maintenance cost and the fuel cost on trucks are relatively higher than those of UAVs. To avoid involving trucks in parcel delivery, researchers have proposed two approaches to increase the delivery area of UAVs. The first one is to locate depots or install charging stations [22, 34, 49] where a UAV can replace or recharge the battery if necessary. The second approach is to explore autonomous mobility such as public transportation vehicles [23, 40]. A UAV can land on the roof of a public transportation vehicle, travel with it, and then leave the vehicle at some position near the customer. If necessary, the UAV can also transfer to another public transportation vehicle. It is easy to see that the first approach requires the supplier to invest in constructing infrastructures, while the second approach is promising to enable a UAV to reach a customer that located far from the depot without much extra cost since the public transportation vehicles are natural mobile platforms traversing our living area.

Since the second approach is more cost-efficient, which may be more favorable to suppliers, we focus on this approach, and we investigate a fundamental problem of how to make use of the public transportation vehicles to transport a UAV from the depot to a customer. Different from the normal delivery vehicles such as trucks, public transportation vehicles are not controlled by logistics companies. They have fixed routes and predefined timetables. Besides, these vehicles cannot precisely follow the timetables. The travel time of a vehicle between two stops is impacted by the departure time (in rush hours or off-peak hours) as well as some uncertainty like traffic congestion, passenger demand, vehicle breakdowns, etc. Therefore, the public transportation network is actually a stochastic time-dependent network. As a delivery UAV may also have limited ability of computation, it is sensible to construct a path before the UAV leaves the depot. Then, the UAV can only memorize some key information about the path and does not need to replan the path during the trip. Therefore, a supplier may wish to construct a reliable path before departure, and we name this problem the reliable UAV path planning problem (RDPP).

Without the term of UAV, the reliable path planning problem in a stochastic time-dependent network is extended from the path planning problem in deterministic non-time-dependent networks [50, 51], stochastic non-time-dependent networks [52], and deterministic time-dependent networks [53, 54]. With uncertainty and time-dependency accounted together, the problem becomes more complex [55, 56]. Some publications have proposed methods for finding the path with the least expected time [57–59]. The path obtained by these approaches can have the least expected travel time, but it is with some risk or, in other words, it is not reliable. A common way to incorporate the travel time variability is to further account for some other metrics such as standard deviation. Then, the reliable path can be redefined as the one that maximizes the probability of guaranteeing a travel time no greater than a given threshold [25, 60–62].

Returning to reliable UAV path planning in a stochastic time-dependent network, we need to model the new problem, which is different from the existing problems in several aspects. The considered stochastic time-dependent network is a combination of the public transportation network and the UAV flight network. We can also regard it as a multimodal network. The node set of this network consists of the vehicle stops and two extra nodes: the depot and the customer. The edge set consists of the links in the public transportation network and some extra links. The extra links connect the two extra nodes with some stop nodes and may also connect two stop nodes if they are not linked yet. The number of extra links should be sufficient such that the two extra nodes are connected in the combined network. This network inherits the stochastic time-dependent feature of the public transportation network. In this network, the UAV should fly by itself and travel with some vehicles, and these two operations should appear alternately on a UAV path. Thus, the UAV may need to wait for the vehicle if it arrives earlier. This is different from many existing publications focusing on road networks where waiting at an intermediate node is not permitted [61].

The traditional shortest path problem in a deterministic non-time-dependent network has been well studied since the proposal of Dijkstra's algorithm [50] and its variations such as A* [51]. Based on these fundamental publications, researchers have proposed approaches to the shortest path problem in a deterministic time-dependent network [53, 54]. The shortest path problem in a network featured with time-dependency and uncertainty attracts lots of attention, and the problem under investigation falls into this category. To the best knowledge of the authors, the problem under investigation has not been considered in the existing literature.

The paper [55] is an early publication discussing the path planning problem under time-dependent uncertainty. Beyond pointing out that treating the stochastic travel times with the corresponding expected values and generating the algorithms for counterpart cases cannot obtain universally valid paths due to the time-dependency, the authors propose a weak form of the consistent condition called stochastic consistency. Some publications have proposed methods for finding the path with the least expected travel time in urban road networks. For example, a label setting algorithm is proposed in [57], a multicriteria A* algorithm is presented in [58], and a bi-level bounds-based label correcting algorithm is discussed in [59]. The path obtained by these approaches can have the least expected travel time, but the path is with risk or, in other words, not reliable.

A common way to incorporate the travel time variability is to add another index to the mean travel time such as the standard deviation. Then, the optimal path can be redefined as the one that maximizes the probability of realizing a travel time no greater than a given threshold [60]. The authors of [61] present a label correcting algorithm to find reliable paths based on several stochastic dominance concepts. Different from the above references focusing on road networks, where waiting is not permitted at any intermediate node, the paper [25] considers the reliable path planning problem in bus networks.

6.3 System Model

Let $\mathcal{G}(\mathcal{N}_1, \mathcal{A}_1, \mathcal{L}_1)$ denote the public transportation network, where $\mathcal{N}_1$ is a set of nodes consisting of the vehicle stops, $\mathcal{A}_1$ is a set of links between pairs of vehicle stops: $\mathcal{A}_1 = \{a_{ij}^l | i, j \in \mathcal{N}_1, l \in \mathcal{L}_1\}$, and $\mathcal{L}_1 = \{l\}$ is the set of vehicle line labels; see Figure 6.3. Let $\mathcal{G}(\mathcal{N}_2, \mathcal{A}_2, \mathcal{L}_2)$ denote a UAV flight network. $\mathcal{N}_2 = \mathcal{N}_1 \cup W \cup C$, where W represents the depot and C represents the customer. We assume that the depot D and the customer C do not locate at any vehicle stops. This is a reasonable assumption because the depot and customers cannot be at any stop in practice, although they may be close to some stops. $\mathcal{A}_2$ consists of some links connecting two nodes in $\mathcal{N}_2$: $\mathcal{A}_2 = \{a_{ij}^0 | i, j \in \mathcal{N}_2\}$, where the label 0 is used to represent UAV flight, i.e. $\mathcal{L}_2 = \{0\}$. They connect the depot D and the customer C with some stop nodes and two stop nodes that are within a certain range but not linked in $\mathcal{A}_1$; see Figure 6.3. The links in this group are called normal links. The second group enables transfers between stops, i.e. the UAV can fly to another stop to catch a vehicle. Let $\mathcal{G}(\mathcal{N}, \mathcal{A}, \mathcal{L})$ be a directed network or graph that combines $\mathcal{G}(\mathcal{N}_1, \mathcal{A}_1, \mathcal{L}_1)$ and $\mathcal{G}(\mathcal{N}_2, \mathcal{A}_2, \mathcal{L}_2)$. Specifically, $\mathcal{N} = \mathcal{N}_2$, $\mathcal{A} = \mathcal{A}_1 \cup \mathcal{A}_2$, and $\mathcal{L} = \mathcal{L}_1 \cup \mathcal{L}_2$. The set $\mathcal{A}_2$ can be added with more UAV flights to make the depot D and the customer C connected in $\mathcal{G}(\mathcal{N}, \mathcal{A}, \mathcal{L})$. Note that if the depot D and the customer C are close to some vehicle lines, some virtual stops (which are the

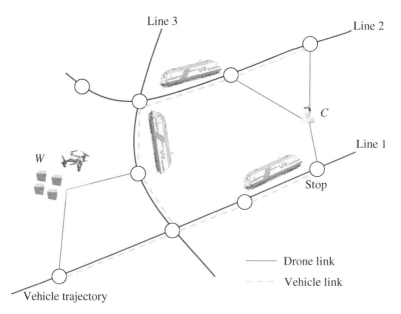

Figure 6.3 An illustration of the public transportation network and the UAV flight network.

positions on some vehicle lines and are closest to the depot D or the customer C) can be added to $\mathcal{N}_1$ at which a vehicle can catch or leave a vehicle without flying too far. In this case, some related links in $\mathcal{A}_1$ are divided into two links. This will increase the number of nodes and links in $\mathcal{G}$ slightly. The links in the second group are called the extra links.

The links in $\mathcal{A}$ are the basis of the system model. Additionally, two other terms also play a pivotal role, i.e. route and path. A route is a series of links that satisfy two conditions: (i) their labels are identical, and (ii) one link and its successor in the series share the same node. The term route is used to describe the travel method between two stops using the same line. Let r_{ij}^l denote a route from node $i \in \mathcal{N}$ to node $j \in \mathcal{N}$ via line $l \in \mathcal{L}$. According to the definition of the route, we can represent a route by $r_{ij}^l = a_{ik_1}^l \oplus a_{k_1 k_2}^l \oplus \cdots \oplus a_{k_{n-1} k_n}^l \oplus a_{k_n j}^l$. Here, $\oplus$ is the concatenation operator that can be used to connect links, routes, and paths. Let $\mathcal{R}$ denote the set of all the routes in $\mathcal{G}(\mathcal{N}, \mathcal{A}, \mathcal{L})$. The term path refers to a series of routes belonging to $\mathcal{R}$. These routes have different labels, and one route needs to share the same node with its successor in the series. Let p_{ij} denote a path from node $i \in \mathcal{N}$ to node $j \in \mathcal{N}$. If there are multiple paths between i and j, we use a superscript for distinction, e.g. p_{ij}^u. In the rest of this chapter, when we mention a path from the depot D to some node j, we omit the starting node. In other words, we use the notation p_j when we refer to a path from D to j.

A UAV can have at most three status:

- Traveling with a public transportation vehicle, i.e. it traverses a normal link.
- Flying, i.e. it traverses an extra link.
- Hovering. This occurs when the UAV arrives at a node earlier than the vehicle on which it plans to ride. Then, it needs to hover at the node to wait for the vehicle.

6.4 One-way Path Planning

In this section, we consider a simple case where a UAV only needs to reach a customer from a depot. We present an approach to plan such a one-way path from the depot to a customer. The round-trip path planning problem will be investigated in the following two sections.

6.4.1 Problem Statement

Before formally stating the problem, we model the traversal time of a path. To do so, we start from the traversal time of a route in $\mathcal{G}(\mathcal{N}, \mathcal{A}, \mathcal{L})$. Let b_{ij}^l denote the traversal time over the route r_{ij}^l. If $l \in \mathcal{L}_1$, traversing this route means that the UAV travels with a public transportation vehicle labeled by l. Although there is a predefined travel time for each route, b_{ij}^l is a random variable considering various traffic factors like congestion. If $l = 0$, it means that the UAV flies by itself. Although the UAV flight is independent of the traffic conditions, b_{ij}^l may still be a random variable if environmental factors such as wind are taken into account. Let $\phi_{b_{ij}^l}(t)$ denote the PDF of b_{ij}^l. For $l \in \mathcal{L}_1$, the PDF can be estimated from empirical observations [63], and for $l = 0$, the data can be collected by conducting filed experiments in different environmental conditions.

Consider two cases where a path has two routes. One case is that the first route is with label 0 and the second route is with a label $l \in \mathcal{L}_1$. The other is that two routes are with the labels $l_1 \in \mathcal{L}_1$ and $l_2 \in \mathcal{L}_1$, respectively, but $l_1 \neq l_2$. The first case means that the UAV flies to a stop and then travels with a vehicle. If the UAV arrives at the stop earlier than the vehicle, the UAV has to wait until the vehicle comes. The second case means that the UAV needs to transfer from one route to another. To do so, the UAV also needs to hover at the intermediate stop for a while. Both lead to some waiting time at the stop. Let c_i^l denote the waiting time at node i before traversing a route that leaves node i and is labeled by l.

Introduce a binary route-path incidence variable x_{ijl}^C, where $x_{ijl}^C = 1$ if the route r_{ij}^l is on the path p_C, and $x_{ijl}^C = 0$ otherwise. Let τ_C be the traversal time of the path p_C, which can be computed as follows:

$$\tau_C = \sum_{r_{ij}^l \in \mathcal{R}} x_{ijl}^C (b_{ij}^l + c_i^l). \tag{6.1}$$

Since b_{ij}^l and c_i^l are random variables, τ_C is also a random variable. Its cumulative distribution function (CDF) is:

$$\Phi_{\tau_C}(t) = \Pr\left(\sum_{r_{ij}^l \in \mathcal{R}} x_{ijl}^C (b_{ij}^l + c_i^l) \leq t\right). \tag{6.2}$$

Let T denote the customer specified delivery time and t_0 be the instant at which the customer makes the order. If the UAV can reach the customer at or before the instant $t_0 + T$, we say the delivery is on time. The probability of fulfilling this requirement is given by $\Phi_{\tau_C}(T)$. We introduce another notation $\Gamma(i)$ to denote the set of successors of node i in the network $\mathcal{G}(\mathcal{N}, \mathcal{A}, \mathcal{L})$. $\Gamma^{-1}(i)$ denotes the set of predecessors of node i.

Definition 6.1   Let P_C denote the set of paths from D to C. We say a path $p_C^u \in P_C$ is T-reliable if the following condition is satisfied [26]:

$$\Phi_{\tau_C^u}(T) \geq \Phi_{\tau_C^v}(T), \quad \forall p_C^v \in P_C, v \neq u. \tag{6.3}$$

Problem Statement: For the given $\mathcal{G}(\mathcal{N}, \mathcal{A}, \mathcal{L})$, T, and t_0, the RDPP problem can be stated as to find a path from the depot D to the customer C such that the probability of delivery on time is maximized. This problem can be formulated following the conventional network flow framework [24]:

$$\max_{x_{ijl}^C} \ \Phi_{\tau_C}(T), \tag{6.4}$$

subject to

$$\sum_{j \in \Gamma(W)} x_{Wjl}^C - \sum_{k \in \Gamma^{-1}(W)} x_{kW\hat{\imath}}^C = 1, \tag{6.5}$$

$$\sum_{j \in \Gamma(C)} x_{Cjl}^C - \sum_{k \in \Gamma^{-1}(C)} x_{kC\hat{\imath}}^C = -1 \tag{6.6}$$

$$\sum_{j \in \Gamma(i)} x_{ijl}^C - \sum_{k \in \Gamma^{-1}(i)} x_{kil}^C = 0, \forall i \neq W, i \neq C, \tag{6.7}$$

$$x_{ijl}^C \in \{0,1\}, \quad r_{ij}^l \in \mathcal{R}, l, \hat{\imath} \in \mathcal{L}, \tag{6.8}$$

The constraints (6.5), (6.6), and (6.7) require the balance of income and outcome of a node. The node D has only one outcome and the node C has only one income. All the other nodes must have either one income and one outcome, or no income and no outcome.

To illustrate the considered problem, we present a small example with two lines and three stops as shown in Figure 6.4. From the depot D, the UAVs may have two paths to reach the customer C. One is $W - i - j - C$ and the other is $W - i - k - C$. If the UAV follows the first path, i.e. the routes r_{Wi}^0, r_{ij}^1, and r_{jC}^0 are selected, the constraints shown in the boxes of Figure 6.4 should hold for nodes D, i, j, and C. Then, the path traversal time is given by $\tau_C = b_{Wi}^0 + c_i^1 + b_{ij}^1 + b_{jC}^0$. If the UAV follows the second path, the corresponding traversal time is given by $\tau_C = b_{Wi}^0 + c_i^2 + b_{ik}^2 + b_{kC}^0$. The better path can be further determined by evaluating the CDF of the path traversal time, i.e. (6.4).

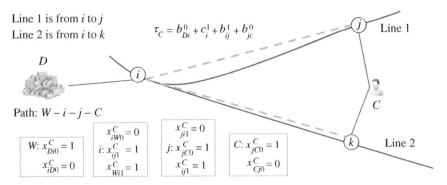

Figure 6.4 A small example to illustrate the path from the depot D to the customer C via nodes i and j.

6.4.2 Proposed Solution

Although the problem (6.4) follows the conventional network flow framework [24], the existing algorithms do not apply to this problem. The reason is that the route traversal times and waiting times are dependent. We show this feature via a simple example in Figure 6.5. The UAV needs to reach node E from node A via node B and C. Route r_{AB} (the solid line connecting nodes A and B) represents a bus route, and its traversal time distribution is shown above the route. Route r_{BC} (the dash line connecting nodes B and C) represents a flight route, and we assume that the traversal time is fixed by neglecting environmental factors such as wind. Route r_{CE} (the solid line connecting nodes C and D) represents a train route that is very reliable. Two trains are running from C to E, and each has the fixed departure instant and traversal time. The UAV departs node A at instant 0 and the arrival instant at node B follows the given distribution. Suppose the UAV arrives at node B before instant 11. After flying for one minute, it arrives at node C before instant 12, which enables it to reach node E at instant 22. However, if the UAV reaches node B after instant 11 but before instant 14, it can reach node C before the next departure of a train. In this case, the UAV needs to hover at node C for some time. It can then arrive at node E at instant 25. It is clear that the waiting time depends on the route traversal time and the vehicle departure instant.

To address the considered problem, we first model the probability density of a path consisting of several routes, and then we discuss how to construct the reliable path.

6.4.2.1 Path Traversal Time

Suppose we have a path p_i from node D to node i. Consider that we extend it by adding a route such that node j is reached. There are three different extensions.

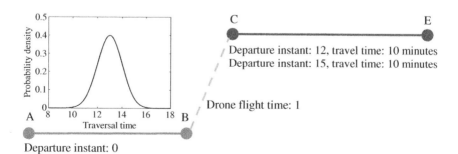

Figure 6.5 An example to illustrate the dependence of travel times and waiting time. The UAV departs A at instant 0 and travels with a bus. When it arrives B, the UAV flies to C. From C, it takes a train to E.

Vehicle route plus UAV flight route: If the end route of p_i is a vehicle route, we can add a UAV flight route a_{ij}^0. The traversal time of path p_i is represented by τ_i, and its PDF is ϕ_{τ_i}. The traversal time of the path p_j is computed by:

$$\tau_j = \tau_i + b_{ij}^0. \tag{6.9}$$

Here, the waiting time at node i is $c_i^0 = 0$ since we always require that if the UAV needs to leave a vehicle, it does so immediately. As the UAV flight has no relationship with how long the UAV travels with a vehicle, the probability densities τ_i and b_{ij}^0 are independent. Then, the PDF of τ_j is given as follows:

$$\phi_{\tau_j} = \phi_{\tau_i} * \phi_{b_{ij}^0}, \tag{6.10}$$

where $*$ is the convolution operator.

Vehicle route plus vehicle route: We can also extend the path p_i ended with a vehicle route by adding another vehicle route a_{ij}^l, $l \in \mathcal{L}_1$. Then, at node i, the UAV may need to hover to wait for the vehicle and the corresponding waiting time is c_i^l. Thus, the traversal time of the path p_j is given by:

$$\tau_j = \tau_i + c_i^l + b_{ij}^l. \tag{6.11}$$

To characterize the waiting time c_i^l, we introduce another random variable d_i^l, which is the departure instant of a vehicle from node i labeled by l. The PDF d_i^l is denoted by $\phi_{d_i^l}$. Similar to that of the route traversal time, the PDF $\phi_{d_i^l}$ can also be extracted from empirical observations. It is easy to understand that the waiting time is the gap between the departure instant of the vehicle with which the UAV plans to travel and the instant at which the UAV arrives at node i, i.e.

$$c_i^l = d_i^l - (\tau_i + t_0). \tag{6.12}$$

Here, $\tau_i + t_0$ is the arrival instant at node i. Clearly, the waiting time at node i depends on two random variables: the UAV arrival instant at node i, i.e. $t_0 + \tau_i$, and the vehicle departure instant from node i, i.e. d_i^l. Summing up the UAV arrival instant and the waiting time at node i, we obtain the UAV departure instant from node i. To further add a vehicle route to the path, it is necessary to know the PDF of the UAV departure instant. Since the UAV departure instant is just the vehicle departure instant, i.e. d_i^l, the PDF of the UAV departure instant is just that of d_i^l. As there may exist multiple vehicles labeled by l to depart from node i, the problem turns to the determination of which vehicle the UAV can catch, see Figure 6.6a. Let H_l denote the number of vehicles serving the line l. The departure instant of the hth (where $h = 1, 2, \ldots, H_l$) vehicle from node i is denoted by d_i^{lh}. Since τ_i and d_i^{lh} are random variables, the computation of the probability of which vehicle the UAV can take is quite complex. We propose the following strategy to simplify the computation.

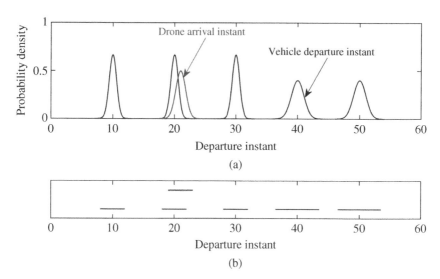

Figure 6.6 (a) There are five vehicles scheduled to depart at instant 10, 20, 30, 40, and 50, respectively. (b) The high probability arrival range (HPAR) of the UAV arrival instant is between the two HPARs of the vehicle departure instants, where $\alpha = 0.005$ and $\beta = 0.995$.

HPAR: Having path p_i, the CDF of the path traversal time τ_i can be computed. Then, we can obtain the CDF of the UAV arrival instant at node i (the distribution is simply shifted by a constant t_0). Introduce a probability interval $[\alpha, \beta]$, where $\alpha, \beta \in (0, 1)$ and $\alpha < \beta$. With the CDF of the UAV arrival instant, we can determine the arrival instants corresponding to the probability α and β, respectively, and these two arrival instants form the HPAR. In addition, for each vehicle serving line l, we do the same for the departure instant. Then, from the HPARs for vehicle departure instants, from the left side, we can find the HPAR whose right-end instant is the closest to the left-end instant of the HPAR of the UAV. We call it the left-range. For example, in Figure 6.6b, the HPAR of the vehicle scheduled to depart at instant 10 is the left-range. Similarly, from the right side, we can find the HPAR whose left-end instant is the closest to the right-end instant of the UAV HPAR. We call it the right-range. In Figure 6.6b, the HPAR of the vehicle scheduled to depart at instant 30 is the right-range. It is possible that there exists some other HPARs between the left-range and the right-range. Here, we take consider the worst case, i.e. the UAV will take the vehicle corresponding to the right-range. The UAV can travel with this vehicle with the probability $\beta - \alpha$. In the example shown in Figure 6.6b, the UAV will take the vehicle scheduled to depart at instant 30. Once we know the vehicle with which the UAV can travel, the PDF of the UAV departure instant is obtained. We use $\phi_{d_i^l|\tau_i+t_0}$ to represent such a PDF. Clearly, it is based on the knowledge of the

HPAR of the random variable $\tau_i + t_0$. Furthermore, we use $\phi_{b^l_{ij} | \tau_i + t_0}$ to represent the PDF of the traversal time b^l_{ij}.

With Eq. (6.12), Eq. (6.11) becomes:

$$\tau_j = d^l_i + b^l_{ij} - t_0. \tag{6.13}$$

We assume that the route traversal time and vehicle departure instant are independent. This is reasonable because for a given scheduled departure instant, the route traversal time depends on the traffic condition mostly. For example, for a vehicle scheduled to depart at 12:00, the actual departure time, say 11:58 or 12:03, has little impact on the travel time to some stop. However, the travel time differs with different scheduled departure instants. For example, the travel time when departing at around 12:00 may be quite different from the one that departs at around 17:00 (which is during the rush hours). Based on this assumption, the PDF of τ_j is computed by:

$$\phi_{\tau_j} = \phi_{d^l_i | \tau_i + t_0} * \phi_{b^l_{ij} | \tau_i + t_0}. \tag{6.14}$$

UAV flight route plus vehicle route: If the path p_i is ended with a UAV flight route, it does not make sense to add another UAV flight route, while it is sensible to add a vehicle route. Then, the situation is the same as the second case discussed above, and the path traversal time and the probability density are given by (6.13) and (6.14), respectively.

Remark 6.1   In the above cases, when the UAV travels with a vehicle, it can turn off the motors to save energy. However, the controller should keep working. The energy consumed by the controller can be ignored compared to that consumed by the motors. Via GPS, the UAV knows its current position. By comparing the GPS position and the precomputed waypoints of the path, the UAV knows when and where to leave the vehicle.

6.4.2.2   Reliable Path Construction

Let $\Phi^{-1}_{\tau_i}(\lambda)$ denote the inverse of the CDF of the traversal time of path p_i:

$$\Phi^{-1}_{\tau_i}(\lambda) = \{\xi | \Pr(\tau_i \le \xi) \ge \lambda\}. \tag{6.15}$$

In words, for a confidence level $\lambda \in (0, 1)$, $\Phi^{-1}_{\tau_i}(\lambda)$ gives the path traversal time that can be achieved with the confidence level λ.

Definition 6.2   Let P_C denote the set of paths from D to C. We say a path $p^u_C \in P_C$ is λ-reliable if the following condition is satisfied [26]:

$$\Phi^{-1}_{\tau^u_C}(\lambda) \le \Phi^{-1}_{\tau^v_C}(\lambda), \forall p^v_C \in P_C, \ v \ne u. \tag{6.16}$$

With Definition 6.2, the original RDPP problem (6.4), can be reformulated as follows:

$$\min_{x_{ijl}^C} \; \Phi_{\tau_C}^{-1}(\lambda), \tag{6.17}$$

subject to (6.5), (6.6), (6.7), and (6.8).

Here, the objective becomes to find the λ-reliable path having the shortest traversal time.

Remark 6.2 Definitions 6.1 and 6.2 are equivalent. Thus, the problem (6.17) is also equivalent with the original problem (6.4).

To address the problem (6.17), we introduce the following assumption and definition.

Assumption 6.1 Following [55], we assume the network $\mathcal{G}(\mathcal{N}, \mathcal{A}, \mathcal{L})$ is stochastic consistent, i.e. the probability of arriving at any node by any given instant cannot be increased by departing later.

Therefore, we always require the UAV to take the first departing vehicle it can catch.

Definition 6.3 Let P_i denote the set of paths from D to i. Consider two paths $p_i^u \in P_i$ and $p_i^v \in P_i$, and $p_i^u \neq p_i^v$. We say p_i^u stochastically dominates p_i^v if the following condition is satisfied:

$$\Phi_{\tau_i^u}^{-1}(\lambda) < \Phi_{\tau_i^v}^{-1}(\lambda), \quad \forall \lambda \in (0, 1). \tag{6.18}$$

This is to say that p_i^u takes shorter time than p_i^v for any confidence level $\lambda \in (0, 1)$. Besides, we say p_i^u is a nondominant path in P_i if it is not dominated by any other path $p_i^v \in P_i$. With Assumption 6.1 and Definition 6.3, we conclude that if p_i^u dominates p_i^v and suppose $j \in \Gamma(i)$, the extended path $p_i^u \oplus r_{ij}^l$ dominates $p_i^v \oplus r_{ij}^l$. Therefore, whenever we encounter a dominated path in the path extension process, it can be discarded safely. This plays a crucial role in our method, because it allows us to prune some candidate paths without evaluating them.

Now, it is the position to present the reliable path construction algorithm. The basic idea is similar to many well-known path planning algorithms for a static graph, i.e. extending the path from the depot D by adding a new route. Obviously, when uncertainty and time-dependency are accounted, some extra operations are needed to evaluate a candidate path.

Let SE be the scan eligible set that stores the nondominant paths and P_i store the nondominant paths from D to i. Initially, SE contains only the path P_W, whose path traversal time probability density is 1. In each iteration, we select one path, say p_i^u, where $i \neq C$, and remove it from SE. Then, for each successor of node i, say $j \in \Gamma(i)$, and each feasible line l leaving node i, we construct an extended path $p_j^u = p_i^u \oplus r_{ij}^l$. To evaluate this extended path, we need to look at the route types of the ending route of the path p_i^u and the added route r_{ij}^l, based on which we compute the PDF of the extended path traversal time by either (6.10) or (6.14). If (6.14) is used, we further need to estimate which vehicle the UAV can take by the HPAR strategy. After knowing the PDF of the traversal time of the extended path p_j^u, we conduct the following two operations to modify the set P_j and SE. Firstly, if p_j^u is not dominated by any other paths in P_j, which is checked according to Definition 6.3, we add it to P_j and SE. Secondly, if any path in P_j and SE is dominated by p_j^u, we remove it from P_j and SE. The algorithm repeats the above process until SE stores paths only to C. These procedures are summarized in Algorithm 6.1. When the algorithm terminates, the node C is associated with a set of nondominant paths and the PDF of these paths have already been computed.

Algorithm 6.1 Constructing the reliable path.

1: Construct an initial path p_W.

2: $SE \leftarrow p_W$.

3: **while** SE has a path to a node i and $i \neq C$ **do**

4:     Select p_i^u $(i \neq C)$ in SE and remove it from SE.

5:     **for** $j \in \Gamma(i)$ and any feasible l **do**

6:         **if** $j \notin p_i^u$ **then**

7:             $p_j^u \leftarrow p_i^u \oplus r_{ij}^l$.

8:             Compute $\phi_{\tau_j^u}$.

9:             **if** p_j^u is not dominated by any path in P_j **then**

10:                 $P_j \leftarrow P_j \cup p_j^u$, $SE \leftarrow SE \cup p_j^u$.

11:                 Remove paths dominated by p_j^u from P_j.

12:                 Remove paths dominated by p_j^u from SE.

13:             **end if**

14:         **end if**

15:     **end for**

16: **end while**

We now analyze the complexity of Algorithm 6.1. Let $|P|$ denote the maximum number of nondominant paths from D to any other node in $\mathcal{N}$, $|\Gamma|$ denote the maximum number of successors of any node, $|L|$ denote the maximum number of lines between two nodes, and $|H|$ denote the maximum number of vehicles serving a line within a period of interest. Clearly, the maximum number of nondominant paths in SE is bounded by $O(|\mathcal{N}||P|)$, which is the maximum number of iterations. For any nondominant path, say p_i^u, the maximum number of extensions is the number of combinations of successors and lines, i.e. $O(|\Gamma||L|)$. For each extended path, say p_j^u, the HPAR strategy requires to compare the HPAR of the extended path with those of at most $O(|H|)$ scheduled departure instants (Line 8 of Algorithm 6.1), and compare this path with at most $O(|P|)$ paths in the set P_j (Line 9 of Algorithm 6.1). Therefore, the overall computational complexity is $O(|\mathcal{N}||\Gamma||L||H||P|^2)$ in the worst case. Clearly, for a given network, $|\Gamma|$, $|L|$, $|H|$, and $|P|$ are fixed, and they should not be large for a practical network. For example, the number of stops connecting directly by some lines to one stop may be a small number if this stop is at a traffic junction, while this number is often two for the normal stops. Besides, for a specific customer, many stops are useless since they may never appear on a feasible path. Removing these nodes from the network significantly reduces the complexity of the proposed approach. Moreover, as we focus on designing the UAV flight before its departure, the computational complexity is not a bottleneck here.

6.4.2.3 Energy-aware Reliable Path

We note that the current problem statement and the proposed reliable path construction algorithm have not considered an important constraint of UAVs, i.e. the limited battery capacity. In this subsection, we extend the proposed approach such that an energy-aware reliable path can be constructed.

We introduce some more notations to model the energy consumption of the UAV. Let E_0 denote the allowed energy to be consumed for the delivery task. Practically, it can be half of the battery capacity. Here, we assume that operations of flying and hovering consume approximately the constant powers, respectively; while when the UAV travels with a vehicle, the energy consumption is assumed to be zero. Recall Remark 6.1, the energy consumption by the controller during this period is ignored. Let q_f and q_h be the consuming powers of the UAV for flying and hovering, respectively. Then, we can relate the energy consumption of the UAV on a path to the corresponding time. Let ϵ^l be the energy consuming power on a route with label l defined as follows:

$$\epsilon^l = \begin{cases} q_f, & \text{if } l = 0, \\ 0, & \text{otherwise.} \end{cases} \tag{6.19}$$

Moreover, let e_i denote the energy consumption on path p_i. Similar to (6.1), e_C can be computed as:

$$e_C = \sum_{r_{ij}^l \in \mathcal{R}} x_{ijl}^C (\epsilon^l b_{ij}^l + q_h c_i^l). \tag{6.20}$$

Obviously, (6.20) is a weighted form of (6.1). Thus, e_C is also a random variable. The CDF of the energy consumption no more than E_0 is given by:

$$\Phi_{e_C}(E_0) = \Pr\left(\sum_{r_{ij}^l \in \mathcal{R}} x_{ijl}^C(\epsilon^l b_{ij}^l + q_h c_i^d) \leq E_0\right).\tag{6.21}$$

Definition 6.4 Let $\eta \in (0, 1)$ be the confidence level for the total energy consumption no more than E_0. We say the path p_C is feasible if the following condition holds:

$$\Phi_{e_C}(E_0) \geq \eta.\tag{6.22}$$

Algorithm 6.2 Constructing the reliable and feasible path.

1: Construct an initial path p_W.
2: $SE \leftarrow p_W$.
3: **while** SE has a path to a node i and $i \neq C$ **do**
4:     Select p_i^u ($i \neq C$) in SE and remove it from SE.
5:     **for** $j \in \Gamma(i)$ and any feasible l **do**
6:         **if** $j \notin p_i^u$ **then**
7:             $p_j^u \leftarrow p_i^u \oplus r_{ij}^l$.
8:             Compute $\phi_{\tau_j^u}$ and $\phi_{e_j^u}$.
9:             **if** $\Phi_{e_C}(E_0) \geq \eta$ **then**
10:                 **if** p_j^u is not dominated by any other path in P_j **then**
11:                     $P_j \leftarrow P_j \cup p_j^u$, $SE \leftarrow SE \cup p_j^u$.
12:                     Remove paths dominated by p_j^u from P_j.
13:                     Remove paths dominated by p_j^u from SE.
14:                 **end if**
15:             **end if**
16:         **end if**
17:     **end for**
18: **end while**

With Definition 6.4, the original problem is extended to maximizing (6.4) subject to (6.5), (6.6), (6.7), (6.8), and (6.22), which is to find the feasible and shortest path in time. The corresponding converted problem becomes maximizing (6.17) subject to (6.5), (6.6), (6.7), (6.8), and (6.22). To address the latter problem, we present Algorithm 6.2, which is extended from Algorithm 6.1 by adding one extra procedure only, i.e. verifying constraint (6.22). Specifically, before verifying whether a path is dominated or not, we check whether this path is feasible; see Line 9. If not, this path is discarded. Such a procedure not only ensures that the stored paths in the set SE are feasible but also reduces the number of iterations for running the algorithm.

6.4.3 Evaluation

A simple network is adopted to illustrate the proposed algorithm. As shown in Figure 6.7, this network consists of 6 nodes and 10 routes. The route labels are shown along each route. The route traversal times are shown in Table 6.1 and the departure instants from four stop nodes are shown in Table 6.2. As all the route traversal times and departure instants are assumed to follow normal distributions in this case, only their mean and standard deviation are provided. The UAV flight routes are more reliable than the vehicle routes, i.e. with smaller deviations. We set $t_0 = 0$, $\alpha = 0.0015$ and $\beta = 0.9985$.

We apply Algorithm 6.1 on the example network, and how it works is shown in Table 6.3. In Iteration 0, SE only contains P_D. In Iteration 1, P_D is the currently considered path. As $\Gamma(W) = \{A, B\}$, we can extend P_D to P_A^1 and P_B^1. Both extensions add UAV flight routes. Since the departure instant is $t_0 = 0$, the path traversal time distributions are just those of the flight routes, shown in Table 6.1. After Iteration 1, $SE = \{P_A^1, P_B^1\}$. In Iteration 2, P_A^1 is extracted from SE. This path can be extended in three ways, as there are three vehicle lines moving out of node A. For line 1, the algorithm first adopts the HPAR strategy to figure out which vehicle the UAV can take. In this case, it is the second vehicle labeled by 1. Then, the path

Figure 6.7 The network used to demonstrate how the proposed algorithm works.

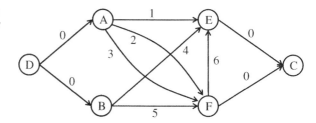

Table 6.1 The distributions of route traversal times.

Route	Traversal time
r_{DA}^0	3(0.1)
r_{DB}^0	5(0.2)
r_{DC}^0	6(0.2)
r_{EC}^0	3(0.3)
r_{AE}^1	23(2.0), 23(2.0), 22(2.0), 21(1.5), 21(1.5)
r_{AF}^2	10(1.2), 10(1.2), 10(1.2)
r_{AF}^3	10(1.0), 10(1.0), 10(1.0), 11(1.2), 11(1.3), 11(1.3)
r_{BE}^4	15(1.4), 15(1.4), 15(1.4), 15(1.4), 15(1.4)
r_{BF}^5	8(1.6), 8(1.6), 8(1.6), 10(1.3), 10(1.3), 10(1.3),
r_{FE}^6	6(1.5), 6(1.5), 6(1.5), 6(1.3), 6(1.3), 7(1.3), 7(1.3)

Table 6.2 The distributions of departure instants.

	Departure instant
d_A^1	$-1(1.0), 9(1.0), 19(1.0), 29(1.0), 39(1.0)$
d_A^2	$5(1.5), 20(1.5), 35(1.5)$
d_A^3	$2(1.1), 8(1.1), 14(1.1), 20(1.2), 26(1.3), 32(1.3)$
d_B^4	$-3(1.4), 7(1.4), 17(1.6), 27(1.6), 37(1.6)$
d_B^5	$-1(1.3), 6(1.3), 13(1.3), 20(1.3), 30(1.4), 40(1.4)$
d_F^6	$2(1.2), 8(1.2), 14(1.2), 20(1.2), 26(1.0), 31(1.0), 36(1.0)$

Table 6.3 The computation process of applying Algorithm 6.1 to the example network.

Iteration	Considered path	Path extension	Traversal time distribution	Vehicle to take	SE
0					$\{P_W\}$
1	P_D	$P_A^1 = P_D \oplus r_{DA}^0$	3(0.1)		$\{P_A^1\}$
		$P_B^1 = P_W \oplus r_{DB}^0$	5(0.2)		$\{P_A^1, P_B^1\}$
2	P_A^1	$P_E^1 = P_A^1 \oplus r_{AE}^1$	32(2.2)	2nd	$\{P_B^1, P_E^1\}$
		$P_F^1 = P_A^1 \oplus r_{AF}^2$	30(1.9)	2nd	$\{P_B^1, P_E^1, P_D^1\}$
		$P_F^2 = P_A^1 \oplus r_{AF}^3$	18(1.5)	2nd	$\{P_B^1, P_E^1, P_F^2\}$
3	P_B^1	$P_E^2 = P_B^1 \oplus r_{BE}^4$	32(1.9)	3rd	$\{P_E^1, P_D^2, P_E^2\}$
		$P_F^3 = P_B^1 \oplus r_{BF}^5$	21(2.1)	3rd	$\{P_E^1, P_D^2, P_E^2, P_F^3\}$
4	P_E^1	$P_C^1 = P_E^1 \oplus r_{EC}^0$	35(2.3)		$\{P_F^2, P_D^2, P_F^3, P_C^1\}$
5	P_F^2	$P_C^2 = P_F^2 \oplus r_{FC}^0$	24(1.5)		$\{P_E^2, P_D^3, P_C^2\}$
		$P_E^3 = P_F^2 \oplus r_{FE}^6$	32(1.6)	5th	$\{P_E^2, P_F^3, P_C^2, P_E^3\}$
6	P_E^2	$P_C^3 = P_E^2 \oplus r_{EC}^0$	35(2.0)		$\{P_F^3, P_C^2, P_E^3\}$
7	P_F^3	$P_E^4 = P_F^3 \oplus r_{FE}^6$	38(1.6)	6th	$\{P_C^2, P_E^3\}$
		$P_C^3 = P_F^3 \oplus r_{FC}^0$	27(2.7)		$\{P_C^2, P_E^3, P_C^3\}$
8	P_E^3	$P_C^4 = P_E^3 \oplus r_{EC}^0$	35(1.7)		$\{P_C^2, P_C^3\}$

traversal time distribution of P_E^1 is computed by Eq. (6.14). Similarly, the path traversal time distributions of P_D^1 and P_D^2 can be obtained, which are 30(1.9) and 18(1.5), respectively. According to Definition 6.3, P_F^2 dominates P_F^1. Thus, after P_F^2 is added to SE, P_F^1 is removed from SE. The algorithm continues until SE contains paths to C only. Therefore, we find two paths P_C^2 and P_C^3 and they are nondominant. To further obtain the whole path, a backtracking method is adopted, and the process of path extension, i.e. the third column in Table 6.3 is used. For example, to construct the whole path of P_C^2, we look for the iteration where it is generated.

It is Iteration 5, and $P_C^2 = P_F^2 \oplus r_{FC}^0$. P_F^2 is generated in Iteration 2 by $P_A^1 \oplus r_{AF}^3$. P_A^1 is generated in Iteration 1 by $P_D \oplus r_{DA}^0$. Thus, $P_C^2 = r_{WA}^0 \oplus r_{AF}^3 \oplus r_{FC}^0$, where P_D can be removed. In words, from the depot, the UAV should first fly to stop A, take line 3 to arrive at stop F, and finally fly to the customer. Using the same backtracking method, we find the whole path for P_C^3, i.e. $P_C^3 = r_{DB}^0 \oplus r_{BF}^5 \oplus r_{FC}^0$. The CDFs of these two paths are shown in Figure 6.8. Clearly, except for some small traversal times, the path P_C^2 is with higher probability than P_C^3 to realize a certain traversal time. For example, following P_C^2, the UAV can reach the customer in 25 minutes with the probability of 0.78, while that of P_C^3 is only 0.25.

Remark 6.3 In the case study, the HPAR strategy is used to determine the vehicle that the UAV can catch. The HPAR corresponds to a very high probability of 0.997 $(\beta - \alpha)$, which is the three-sigma rule for normal distributions. If we relax this probability a bit, different paths with higher risk of not being realized can be found.

Moreover, we take into account the energy consumption of the UAV and apply Algorithm 6.2 to the example network. Here, $E_0 = 20$, $q_f = 1$, $q_h = 1$, and $\eta = 0.99$, and all the other parameters are the same as above. The computation process is summarized in Table 6.4. As an additional constraint (6.22) is considered, there are some differences between Tables 6.4 and 6.3. In particular, when path P_D^1 is constructed, the energy consumption distribution shown in the fifth column cannot satisfy (6.22). Thus, different from Table 6.3, this path is not added into the set of SE. This case occurs several times; see the paths P_E^2, P_E^3, P_E^4, and P_C^3. So,

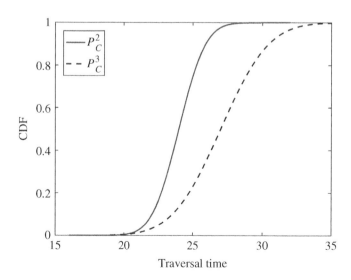

Figure 6.8 The CDFs of the two found paths.

Table 6.4 The computation process of applying Algorithm 6.2 to the example network.

Iteration	Considered path	Path extension	Traversal time distribution	Energy consumption distribution	Vehicle to take	SE
0						$\{P_W\}$
1	P_D	$P_A^1 = P_D \oplus r_{DA}^0$	3(0.1)	3(0.1)		$\{P_A^1\}$
		$P_B^1 = P_W \oplus r_{DB}^0$	5(0.2)	5(0.2)		$\{P_A^1, P_B^1\}$
2	P_A^1	$P_E^1 = P_A^1 \oplus r_{AE}^1$	32(2.2)	9(1.0)	2nd	$\{P_B^1, P_E^1\}$
		$P_F^1 = P_A^1 \oplus r_{AF}^2$	30(1.9)	20(1.5)	2nd	$\{P_B^1, P_E^1\}$
		$P_F^2 = P_A^1 \oplus r_{AF}^3$	18(1.5)	8(1.1)	2nd	$\{P_B^1, P_E^1, P_D^2\}$
3	P_B^1	$P_E^2 = P_B^1 \oplus r_{BE}^4$	32(1.9)	17(1.6)	3rd	$\{P_E^1, P_F^2\}$
		$P_F^3 = P_B^1 \oplus r_{BF}^5$	21(2.1)	13(1.3)	3rd	$\{P_E^1, P_F^2, P_F^3\}$
4	P_E^1	$P_C^1 = P_E^1 \oplus r_{EC}^0$	35(2.3)	12(1.0)		$\{P_F^2, P_F^3, P_C^1\}$
5	P_F^2	$P_C^2 = P_F^2 \oplus r_{FC}^0$	24(1.5)	14(1.1)		$\{P_F^3, P_C^2\}$
		$P_E^3 = P_D^2 \oplus r_{FE}^6$	32(1.6)	16(2.1)	5th	$\{P_D^3, P_C^2\}$
6	P_F^3	$P_E^4 = P_F^3 \oplus r_{FE}^6$	38(1.6)	23(2.7)	6th	$\{P_C^2\}$
		$P_C^3 = P_F^3 \oplus r_{FC}^0$	27(2.7)	19(1.3)		$\{P_C^2\}$

Algorithm 6.2 completes in fewer iterations. Finally, there is only one returned path, i.e. P_C^2, which is the feasible and shortest path in time.

6.5 Round-trip Path Planning in a Deterministic Network

Beyond the one-way path planning discussed in Section 6.4, in the section and the following one, we discuss the round-trip path planning problem. From simple to complex, we discuss the round-trip path planning problem in a deterministic network in this section. In Section 6.6, we will discuss the round-trip path planning problem in a stochastic network.

6.5.1 Deterministic Model

There may exist several routes between two nodes, and the departure time as well as the arrival time of each route are given in a predefined timetable. In general, the travel time between two nodes by a route in the peak hours is longer than that in the off-peak hours. For instance, in Figure 6.9b, it takes service-1 10 minutes to move from stop 1 to stop 2, while it takes service-3 15 minutes. At a stop, a

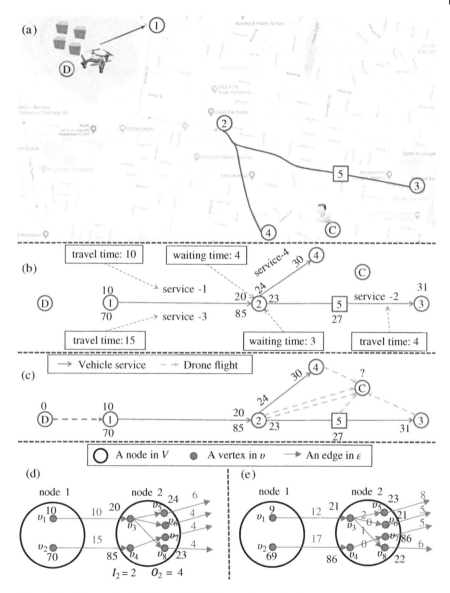

Figure 6.9 Converting $G(V, E)$ to $\mathcal{G}(\mathcal{V}, \mathcal{E})$ by turning nodes into vertices and complex edges into simple edges. The black numbers near nodes and vertices are time instants, and the red numbers near edges represent energy consumption. Node D represents the depot, and node C represents the customer.

service-to-service transfer is allowed, and the corresponding waiting time depends on the arrival time of the incoming service and the departure time of the outgoing service, see Figure 6.9b.

To enable a UAV to depart from and return to a service when the service is moving, some points are inserted into two stops linked by the service. This will reduce the UAV flight time, especially when the stops are far away from each other. For example, in Figure 6.9a, if the UAV takes service-2, it is better for the UAV to leave service-2 at node 5 and then flies to the customer C. The time passing by these points can be estimated according to the departure time and arrival time at the two stops. Let V denote the union of the inserted points, the stops, the depot, and the customer, and an element of V is called a node. Let E be an edge set. Many nodes in V are linked by services. We now add more edges to E representing UAV flights. Let d_0 be a given constant, representing a distance threshold. Let the function $d(i,j)$ compute the distance between nodes i and j, where $i, j \in V$. If $d(i,j) \leq d_0$ and there is not a service from i to j, a new edge (i,j) is added to E. In Figure 6.9b, although the distance between node 5 and 3 may be smaller than d_0, we do not add an edge for them, because there is a service from node 5 to node 3. It is worth pointing out that more than one edge may be added between two nodes. For instance, two edges are added between node 2 and node C in Figure 6.9c, since there are two services (i.e. service-1 and service-3) the UAV can take to reach node 2. Now, the multimodal network $G(V, E)$ consists of the public transportation network and UAV flights.

Remark 6.4 $G(V, E)$ also depends on when the UAV plans to depart from the depot. For instant, suppose the UAV can depart from the depot at time instant 60 in the example of Figure 6.9c. Then, only service-3 will be kept in this case, while all the other services will be unavailable to the UAV.

6.5.1.1   Extended Multimodal Network

$G(V, E)$ is a complex network, as there may exist several edges in E linking two nodes in V. We construct a simple network $\mathcal{G}(\mathcal{V}, \mathcal{E})$ from $G(V, E)$. We call an element in V a node and an element in $\mathcal{V}$ a vertex for distinction. We require that there is at most one edge in $\mathcal{E}$ linking two vertices in $\mathcal{V}$. The construction procedures are detailed as follows.

1. Generate vertices. For every node $i \in V$, there is a certain number I_i of incoming edges and also a number O_i of outgoing edges. Then, node $i \in V$ turns into $I_i + O_i$ vertices in $\mathcal{V}$. For instance, in Figure 6.9c, node 2 has two incoming edges and 4 outgoing edges. Thus, in Figure 6.9d, node 2 turns to be 6 vertices. Note that the generated vertices from a node in V are associated with the same label (by which we can inquire the physical location of the vertex). Any vertex other than those relating to UAV flights is now with a time instant, which is known from the timetable of the public transportation vehicles. For instance, v_5 in Figure 6.9d is with the time 24, which is obtained from the timetable of

service-4. v_6 has not been assigned with any time instant yet since it relates to a UAV flight, and we discuss this below.

2. Convert a node–node edge to vertex–vertex edges. With the generated vertices, an edge between two nodes in V turns into several different edges connecting the right vertices. For instance, in Figure 6.9c, node 1 and node 2 are connected by one edge, and this edge turns into two edges in Figure 6.9d. By this procedure, a complex edge in E is converted to several simple edges in $\mathcal{E}$. Moreover, let ϵ_s, ϵ_h, and ϵ_f denote the energy consumption rates for traveling with a service, hovering and flying, respectively [64]. Then, a generated vertex–vertex edge has the corresponding energy consumption, given the time needed to complete the edges by the UAV. For traveling with a service, the time to complete the edge can be retrieved from the timetable. For flying, we use the length of the edge and the average UAV speed v to compute the needed time.

3. Generate inter-node edges. The inter-node edges are generated within a certain node in V to represent the hovering, the take-off, and the landing actions. The number of inter-node edges depends on the number of the generated vertices from this node in Procedure 1 and the time instants associated with these vertices. We sort these instants and link two vertices by their respective instants. For instance, in Figure 6.9d, v_3 is linked with v_5 and v_8, since the instant associated with v_3 is smaller than those of v_5 and v_8. v_3 is also linked with v_6, and this edge represents the take-off action. The number of the generated inter-node edges is at most $I_i + O_i$.

4. Modify the instant on vertices and the energy consumption on edges. Let t_t and t_l denote the time needed for take-off and landing, respectively, and e_t and e_l be the corresponding energy consumption, respectively. Then, any vertex corresponding to an incoming service has a time annotation that is t_t later than the service arrival, and an additional energy consumption e_t is added to the edge. Any vertex corresponding to an outgoing service has a time annotation that is t_l earlier than the service departure, and an additional energy consumption e_l is added to the edge. For a vertex representing an intermediate service node, it is linked to a vertex by a service and another vertex by a UAV flight. If the service is the incoming service, then this vertex has a time annotation that is t_t later than the service arrival. If the service is the outgoing service, then this vertex has a time annotation that is t_l earlier than the departure time. Also, the corresponding energy for take-off or landing is added to the edge. Figure 6.9e shows the instants on vertices and the energy consumption on edges modified from Figure 6.9d, given $t_t = 1, t_l = 1, e_t = 1, e_l = 1, \epsilon_s = 1, \epsilon_h = 1$, and $\epsilon_f = 1$. For example, the instant with v_3 becomes 21 in Figure 6.9e, and the instant with v_8 becomes 22. Let t_d and e_d denote the time and energy needed for dropping off the parcel at the customer. An edge reaching the customer vertex has an extra energy consumption of $\frac{e_d}{2}$, and the arrival instant is extended by $\frac{t_d}{2}$. An edge leaving the customer vertex has an extra energy consumption of $\frac{e_d}{2}$, and the departure instant is extended by $\frac{t_d}{2}$. The vertices corresponding to a UAV flight can now be assigned with the right departure and

arrival instants and so as the energy consumption. For example, v_6 is with the instant of 22, and the edge connected with v_6 has an energy consumption of 5, given $e_d = 2$. Note that since we split the energy consumption for dropping off the parcel to the corresponding edges representing UAV flights, an edge linking two vertices with label C has zero energy consumption. Moreover, the energy consumption on an inter-node edge can be computed, given the UAV status and the time instants of the two vertices. For instance, the inter-node edge between v_3 and v_8 is with the energy consumption of 1.

With the above procedures, we obtain a simple and directed graph $\mathcal{G}(\mathcal{V}, \mathcal{E})$. In $\mathcal{G}(\mathcal{V}, \mathcal{E})$, each vertex is associated with a time instant and a label, the vertex–vertex connection is determined by their associated time instants, and the cost of each edge is the energy consumption. To facilitate the problem formulation in the subsequent section, we need some new symbols. Let ϕ_i denote the label of vertex $i \in \mathcal{V}$, α_i denote the time instant associated with the vertex $i \in \mathcal{V}$, and c_{ij} denote the cost (i.e. energy consumption) of the edge $(i, j) \in \mathcal{E}$.

6.5.2 Problem Statement

6.5.2.1 Shortest UAV Path Problem

We look for the shortest path in terms of time from the depot to the customer and then back to the depot in the graph $\mathcal{G}(\mathcal{V}, \mathcal{E})$, such that the UAV does not violate the energy constraint. This path is called the shortest UAV path.

Let P_1 and P_2 denote the depot-to-customer path and the customer-to-depot path, respectively. P_1 starts from a vertex with the label of D and ends at a vertex with the label of C, while P_2 starts from a vertex with the label of C and ends at a vertex with the label of D. Thus, P_1 and P_2 form the whole UAV path. Let $\mathcal{V}_1$ and $\mathcal{V}_2$ be two duplicated the vertex sets of $\mathcal{V}$. Then, P_1 should consist of a sequence of vertices in $\mathcal{V}_1$, and P_2 should consist of a sequence of vertices in $\mathcal{V}_2$. Suppose P_1 has n vertices and P_2 has m vertices. Let $P_1 = \{v_1, v_2, \dots, v_n\} \in \mathcal{V}_1 \times \dots \times \mathcal{V}_1$ and $P_2 = \{v_{n+1}, v_{n+2}, \dots, v_{n+m}\} \in \mathcal{V}_2 \times \dots \times \mathcal{V}_2$. So, the whole path $P = \{v_1, \dots, v_n, v_{n+1}, \dots, v_{n+m}\}$. Note that any two adjacent vertices in P are linked by an edge, i.e. $(v_i, v_{i+1}) \in \mathcal{E}$. The numbering i is the position in the sequence and needs not to relate to any canonical labeling of the vertices. Let E_0 be the initial energy of the UAV.

Then, the shortest UAV path problem is formulated as:

$$\min_{P} \alpha_{v_{n+m}} \tag{6.23}$$

subject to

$$\phi_{v_1} = \phi_{v_{n+m}} = D, \tag{6.24}$$

$$\phi_{v_n} = \phi_{v_{n+1}} = C, \tag{6.25}$$

$$(v_i, v_{i+1}) \in \mathcal{E}, \forall 1 \leq i < n + m, \tag{6.26}$$

$$\sum_{i=1}^{n+m-1} c_{v_i, v_{i+1}} \leq E_0. \tag{6.27}$$

The objective function (6.23) minimizes the time instant the UAV returns to the depot D. Constraint (6.24) specifies that the first and final vertices should be with the label D. Constraint (6.25) requires that two vertices with the label C should be on the path. Constraint (6.26) requires that any two adjacent vertices on the path are linked by an edge. Constraint (6.27) requires that the battery constraint does not violate.

6.5.3 Proposed Solution

In this section, we first present our solution to construct the UAV path P in $\mathcal{G}(\mathcal{V}, \mathcal{E})$. Then, we extend the method to deal with the uncertainty in $\mathcal{G}(\mathcal{V}, \mathcal{E})$.

6.5.3.1 The Dijkstra-based Algorithm

In this subsection, we present a Dijkstra-based algorithm to address the problem (6.23). Dijkstra's algorithm is a well-known algorithm to compute the shortest path between a source and a destination in a graph. The basic idea of Dijkstra's algorithm is as follows. Initially, all the vertices in the graph are marked as unvisited. Every vertex is assigned to a tentative distance value: zero for the source vertex and infinity for all other vertices. The source is set as the current vertex. Consider all the unvisited vertices linked to the current vertex and calculate their tentative distances through the current vertex. For each of these unvisited vertices, if the newly calculated tentative distance is smaller than its current value, the tentative distance is updated. Mark the current vertex as visited. Continue this operation by setting the unvisited vertex with the smallest tentative distance as the current vertex. The algorithm terminates when the destination vertex is marked visited (i.e. the shortest path is found) or the smallest tentative distance among the unvisited vertices is infinity (i.e. there is not such a path). Moreover, a straightforward extension of this Dijkstra's algorithm is to produce a shortest-path tree from the source to multiple destinations. This can be achieved by slightly modifying the termination conditions to: the algorithm terminates when all the destination vertices are marked visited or the smallest tentative distance among the unvisited vertices is infinity.

There are some differences between the conventional "shortest path problem" and the problem (6.23). Firstly, in the conventional problem, an edge is with a length, while in our problem, an edge is with a cost of energy consumption. Additionally, there is a given pair of source and destination in the conventional "shortest path problem". However, in our problem, as we duplicate a node in V

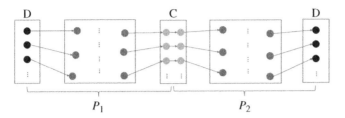

Figure 6.10 Constructing the shortest UAV path.

to get several vertices in $\mathcal{V}$, there may exist multiple vertices corresponding to the depot node and customer node. Therefore, there are multiple pairs of "source and destination" in $\mathcal{G}(\mathcal{V}, \mathcal{E})$.

The proposed method, as presented in Algorithm 6.3, is based on aforementioned Dijkstra's variant. As shown in Figure 6.10 and as mentioned previously, the whole path P generated by Algorithm 6.3 consists of P_1 and P_2. We first consider the construction of P_1, see Line 1–9 in Algorithm 6.3. For each vertex with the label of D in $\mathcal{V}_1$, we apply the Dijkstra's variant to construct the minimum-energy tree by taking the vertices with the label of C as destinations. As there may be multiple vertices with the label of D, there may exist multiple different paths to a certain vertex with the label of C. Any vertex with the label of C is marked by the minimum energy consumption among these paths. So far, we have constructed a set of feasible paths for P_1, and these feasible paths may have different source vertices (with label D) and destination vertices (with label C). Since we split the cost on a C–C edge to the incoming and outgoing edges (see Procedure 4), the tentative costs of the C vertices having outgoing edges directly take the tentative costs of the corresponding C vertices having incoming edges. Now, we continue to construct the feasible paths for P_2, see Line 11–18 in Algorithm 6.3. The operation is similar to what we have done. Specifically, for each vertex with the label of C (whose tentative cost is obtained from the corresponding destination vertex of the feasible paths for P_1), we apply the Dijkstra's variant again to construct the minimum-energy tree by taking the vertices with the label of D in $\mathcal{V}_2$ as destinations. Again, there may exist multiple paths reaching a certain vertex with the label of D, and we only store the information corresponding to the one with the minimum energy consumption. Now, we obtain a set of feasible paths for P, and they may have different destination vertices with the label of D. Among these paths, we choose the one with the following properties: the ending vertex (with label D) is with the smallest instant and the overall energy consumption is no larger than E_0.

It is worth pointing out that the information of E_0 can be used in the algorithm. Specifically, whenever a vertex is to be assigned to a tentative costs over E_0, the tentative cost is set as infinity. This can reduce some unnecessary calculations, because the paths via this vertex are infeasible.

Algorithm 6.3 The shortest UAV path construction.

1: Initialization: mark the vertices in $\mathcal{V}_1$ and $\mathcal{V}_2$ as unvisited. Assign a tentative cost zero to the D vertices in $\mathcal{V}_1$ and infinity to the others vertices in $\mathcal{V}_1$ and $\mathcal{V}_2$.

2: **for** each D vertex in $\mathcal{V}_1$ **do**

3:     Set this vertex as the current vertex.

4:     **while** any C vertex in $\mathcal{V}_1$ that has an incoming edge is unvisited, or the smallest tentative cost of the unvisited vertices in $\mathcal{V}_1$ is smaller than infinity **do**

5:         Consider the unvisited vertices in $\mathcal{V}_1$ linked to the current vertex and calculate their tentative costs through the current vertex.

6:         For each of these unvisited vertices, if the new tentative cost is smaller than its assigned value, the tentative cost is updated. Mark the current vertex as visited.

7:         Set the unvisited vertex in $\mathcal{V}_1$ with the smallest tentative cost as the current vertex.

8:     **end while**

9: **end for**

10: Assign the tentative costs of the C vertices in $\mathcal{V}_1$ having incoming edges to the connected C vertices in $\mathcal{V}_2$ having outgoing edges.

11: **for** each C vertex in $\mathcal{V}_2$ that has an outgoing edge **do**

12:     Set this vertex as the current vertex.

13:     **while** any D vertex in $\mathcal{V}_2$ that has an incoming edge is unvisited, or the smallest cost of the unvisited vertices in $\mathcal{V}_2$ is smaller than infinity **do**

14:         Consider the unvisited vertices in $\mathcal{V}_2$ linked to the current vertex and calculate their tentative costs through the current vertex.

15:         For each of these unvisited vertices, if the new tentative cost is smaller than its assigned value, the tentative cost is updated. Mark the current vertex as visited.

16:         Set the unvisited vertex in $\mathcal{V}_2$ with the smallest tentative cost as the current vertex.

17:     **end while**

18: **end for**

6.5.3.2 Reliable UAV Path

In practice, the public transportation network may not operate exactly as scheduled. Specifically, a public transportation vehicle may arrive at a stop earlier or later than the scheduled time. Thus, although the UAV flight is independent on the public transportation network and the flight can be controlled, the delivery may fail if the uncertainty of the public transportation network is not taken into account. The delivery failure can occur in cases where the UAV fails to "take" the scheduled vehicle or the total energy consumption following the designed path is larger than E_0. In this subsection, we discuss how to construct a reliable path for the UAV.

Failing to "take" the scheduled vehicle occurs when the UAV arrives at the stop later than the vehicle departure time, see Figure 6.11. If the UAV arrives at a stop earlier than the vehicle, more energy is to be consumed, which may result in that the total energy consumption is over E_0. Thus, a reliable UAV path should ensure

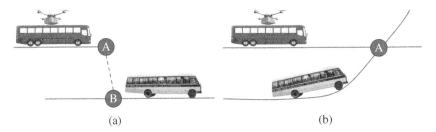

(a)	(b)

Figure 6.11 Delivery fails when the bottom vehicle leaves the stop earlier than the arrival of the UAV.

that the UAV can arrive at a stop earlier than the scheduled vehicle, and the energy consumption is no more than E_0.

Our approach to improve the reliability of the UAV path is to ensure that the constructed edges in $\mathcal{E}$ are valid with a sufficient large probability under the expected inaccuracy of the timetables. Moreover, the costs of the edges should be assigned with reasonable larger values to accommodate the expected longer waiting times at stops. To achieve these, we take into account the distribution of a vehicle's appearance time at a stop, which can be collected by simply installing a GPS module on the vehicle. Let ρ_0 be a given constant that is close to 0. From the distribution of a vehicle's appearance time at a stop, we can easily find two instants: the first one, denoted by t_1, is the instant such that the probability that the vehicle arrives at the stop before t_1 is no more than ρ_0; and the second one, denoted by t_2, is the instant such that the probability that the vehicle departs from the stop after t_2 is no greater than ρ_0; see Figure 6.12. Then, the probability that the vehicle appears at the stop between t_1 and t_2 is $1 - 2\rho_0$. For simplicity, the uncertainty of the UAV flights such as the wind effect is not considered.

For all the public transportation vehicles, we can obtain their corresponding t_1 and t_2, indicating the vehicle's appearance at a stop with high probability. We make the following modifications to the construction of $\mathcal{G}(\mathcal{V}, \mathcal{E})$. Firstly, for any vertex that has an outgoing service, it is associated with t_1 corresponding to that service; and for any vertex that has an incoming service, it is associated with t_2

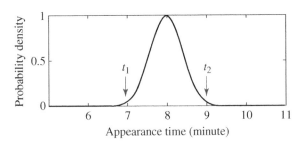

Figure 6.12 The distribution of the appearance time of a vehicle at a stop.

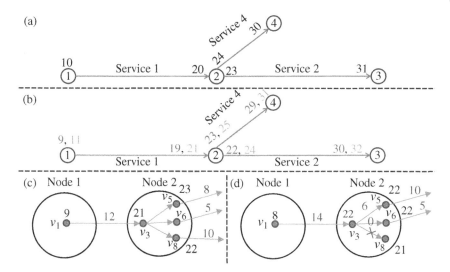

Figure 6.13 Converting $G(V, E)$ to $\mathcal{G}(\mathcal{V}, \mathcal{E})$ accounting for uncertainty. (a) The original network. (b) The values of t_1 (gray) and t_2 (light gray), which are 1 before or after the scheduled time shown in (a), respectively. (c,d) Results of applying the modified procedures.

corresponding to that service. In other words, for a vehicle's departure time, we use the "lower bound" t_1; while, for a vehicle's arrival time, we use the "upper bound" t_2. This setting can improve the reliability of the transfer between different vehicles. An instance is shown in Figure 6.13, where t_1 and t_2 are 1 before and after the scheduled times for all the vehicles. Specifically, v_3 takes the instant of 21 (t_2), and v_8 takes the instant of 22 in Figure 6.13c. After the instant modification procedure, v_3 takes the instant of 22 (with the additional time for take-off), and v_8 takes the instant of 21 (with the additional time for landing) in Figure 6.13d. In this instance, v_3 and v_8 are not connected. Thus, by improving the reliability of transfer, the number of edges in $\mathcal{V}$ is reduced. Besides the modification on the instants associated with the vertices, the cost of the edges are also modified. We assign the cost of an edge to tackle the worst case. Specifically, the node–node edges will be with the cost calculated based on the new instants, and inter-node edges will be with the cost calculated based on the longest waiting time at the node. For example, the cost of the edge between v_1 and v_3 in Figure 6.13d becomes 14, which consists of a landing action, traveling with the service for the time of 12, and a take-off action. The edge between v_3 and v_5 has the cost of 6 to accommodate the case that service-1 arrives at node 2 at instant 19, while service-4 departs at instant 25. Via these modifications, the edges of $\mathcal{E}$ become reliable, and the path generated by Algorithm 6.3 is tolerant to the expected network uncertainty.

6.5.3.3 Extended Coverage

In this subsection, we investigate the gain in terms of the extended coverage area. In the proposed approach, thanks to the vehicles' transportation, the UAV can reach some position with much less energy consumption than flying only. Consider the two means for the UAV to reach the vertex $i \in V$. By flying only, the UAV has to consume $\epsilon_t t_t + \epsilon_f \frac{d(i,D)}{v}$ amount of energy, where the first term is for take-off and the second term is for flying. Let $\varepsilon(i)$ denote the residual energy when the UAV reaches the vertex i. If the following condition holds

$$E_0 - \varepsilon(i) < \epsilon_t t_t + \epsilon_f \frac{d(i,D)}{v}, \tag{6.28}$$

it means that reaching the vertex i by the proposed method consumes less energy than the way of flying only. Thus, the coverage area can be extended from the location of vertex i.

We present a lower bound of the extended coverage area. The model here is based on one possible schedule: the UAV first takes the public transportation network to reach some position; from this position, the UAV flies by itself to the customer; and finally, the UAV flies back to the depot. Now, let C denote a feasible customer. Following the above schedule, the location of C needs to satisfy the following condition:

$$\varepsilon(i) \geq \epsilon_f \frac{d(i,C) + d(C,D)}{v} + e_d + e_l. \tag{6.29}$$

Clearly, the right-hand side of (6.29) is the energy required to finish the trip from vertex i to C and then from C to D. From (6.29), we can easily obtain:

$$d(i,C) + d(C,D) \leq \frac{v}{\epsilon_f}(\varepsilon(i) - e_d - e_l). \tag{6.30}$$

In other words, (6.30) indicates that the position of C falls into an ellipse with the positions of D and vertex i as the foci.

Let $C(i)$ denote the area of the ellipse specified by (6.30), then $C(i) = \{C \mid C$ satisfies (6.30)$\}$. Therefore, the total coverage is given by $\cup_{i \in V} C(i)$. By removing the overlapping area with the coverage region by the UAV directly flight (which is in general a disk centered at the depot D), we can obtain the newly coverage area brought by the public transportation network.

6.5.4 Evaluation

In this section, we first demonstrate how the proposed method works. We consider a network as shown in Figure 6.14a. There are 5 stops (U, V, W, X, and Y) and 8 services (S1–S8 marked along the corresponding edges). The departure and arrival times of these services are given (see the light gray numbers besides the stops). Some UAV flights (undirected) are added, which are represented by the black lines.

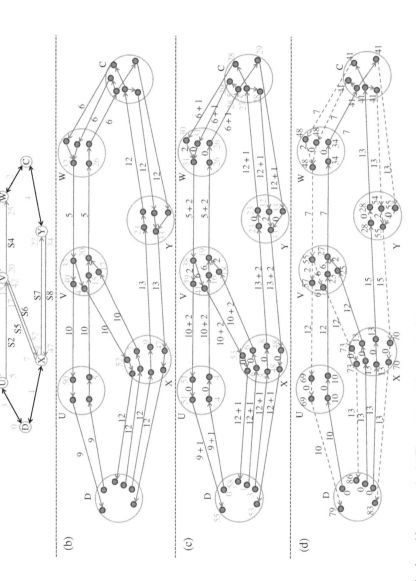

Figure 6.14 A working example. (a) The multimodal network with the depot, the customer, five stops, some services, and some UAV flights. The light gray numbers on the service edges are either an arrival or departure time instant. The light gray numbers on the UAV flight edges are the flight time. (b) The graph obtained by Procedures 1–3 discussed in Section 6.5.1. Some time instants are inherited from the service arrival and departure times. The red numbers on the edges represent the energy consumption of the UAV. (c) The graph after applying Procedure 4 on the graph of (b). The instants of vertices and the energy consumption on edges are modified. (d) The results of applying Algorithm 6.3. The dark gray numbers indicate the minimum accumulated energy consumption to reach the vertices from the depot.

The numbers on these added edges are the time for the UAV to complete the flights. The delivery request is activated at time instant 0. The parameters are set as follows: $\epsilon_s = 1$, $\epsilon_h = 2$, $\epsilon_f = 3$, $t_t = 1$, $t_l = 1$, $t_d = 1$, $e_t = 1$, $e_l = 1$, and $e_d = 2$.

By applying the Procedures 1–3, a graph with simple edges is obtained as shown in Figure 6.14b. Some vertices relating to vehicle services have been marked with time instants, representing either a vehicle arrival time or departure time; see the light gray numbers besides vertices in Figure 6.14b. Additionally, numbers are assigned to edges indicating the energy consumption of the UAV; see the red numbers along the edges in Figure 6.14b. After applying Procedure 4, a complete graph is obtained as shown in Figure 6.14c, where the properties of vertices and edges have been modified.

The proposed algorithm, i.e. Algorithm 6.3, is then applied in the complete graph, and the results are shown in Figure 6.14d. The purple numbers near the vertices are the minimum accumulated energy consumption the UAV needs to spend to reach these vertices. By a standard backtracking method, the paths representing by a sequence of nodes (or vertices) can be identified. For instant, we have a look at the path leading to the minimum energy consumption (i.e. 79) among all the feasible paths. Firstly, the node corresponding to the vertex having the energy consumption 79 is added to the end of the path, i.e. $P = \{D\}$. Looking backwards, the nodes before D should be U, V, and W, i.e. $P = \{W, V, U, D\}$. At node W, we need to determine which vertex leads to the considered path. By accounting for the energy consumption of the vertices with the label of W, we know it is the second vertex on the right-hand side. Then, the nodes C, W and V are added to the path, i.e. $P = \{V, W, C, W, V, U, D\}$. At node V, we need to make another decision as there are two possible paths. Again, by accounting for the energy consumption of the vertices with the label of V, we know that node X should be added to the path. The next node added to the path is node D. Thus, the whole path is $P_{79}^{55} = \{D, X, V, W, C, W, V, U, D\}$. Via this path, the energy consumption is 79 and the return instant is 55. Similarly, we can construct the path that has the energy consumption of 83 and the return instant of 53, i.e. $P_{83}^{53} = \{D, X, V, W, C, Y, X, D\}$; as well as the path that has the energy consumption of 86 and the return instant of 58, i.e. $P_{86}^{58} = \{D, X, V, W, C, W, V, X, D\}$. If $E_0 = 90$, then all the three paths are feasible and the best one has the return instant of 53. If $E_0 = 80$, there is only one feasible path, and the return instant is 55.

Now, we demonstrate the extended coverage area achieved by the proposed method. To make the results easy to understand, we consider a network shown in Figure 6.15. Service-1 (S1) schedules to depart from node 1 (N1) at 10 and reach node 2 (N2) at 15. Service-2 (S2) schedules to depart from node 2 (N2) at 18 and arrive at node 3 (N3) at 21. The initial energy E_0 is set as 80, the UAV average speed is set as $v = 0.5$, and all the other parameters are the same as above.

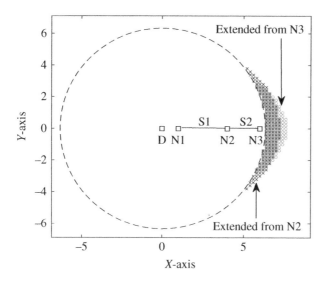

Figure 6.15 The extended coverage from nodes 2 and 3.

Applying the procedures discussed in Section 6.5.1.1, we can compute the residual energy if the UAV "takes" the services to reach nodes 2 and 3, which are 66 and 59, respectively. The residual energy satisfies the condition (6.28). Thus, some extended coverage can be achieved from nodes 2 and 3. By (6.30), we can compute the extended coverage area from these two nodes, which are shown in Figure 6.15. The overall extension is about 8% of the coverage area achieved by flying only (a disk centred at the depot D).

Moreover, we analyze the impact of traffic freeze. Traffic freezes on some parts of roads happen when accidents occur such as collisions or break-down. This significantly impacts the performance of the proposed parcel delivery system, which depends on the transportation network to get to the customers far from the depot. When the traffic of some road freezes, some transportation service on the affected road cannot transport UAVs. Here, we consider a network shown in Figure 6.16a, where the timetables of the services are also shown. In the results shown in Figure 6.16, the coverage area by UAV flight only is regarded as 1. With the assistant of the transportation network, the coverage increases. Figure 6.16b shows that when all the services function well, the extended coverage is increased by 30%. Such an extension is achievable only at some particular time. The reason is that the UAV can "take" a service only at some particular time instant, such as 10. In this case, the UAV can "take" service-1, transfer at node 2, and "take" service-2 to reach the farthest point from the depot. After delivery, it needs to catch service-3, transfer to service-4 and returns to the depot. Figure 6.16c

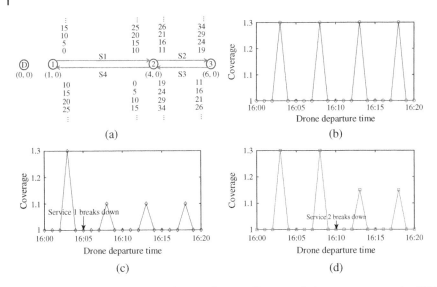

Figure 6.16 (a) A network with four services. (b) The extended coverage versus the UAV departure time when all the services operate normally. (c) Service-1 breaks down at 16:05. (d) Service-2 breaks down at 16:10.

shows the extended coverage when service 1 breaks down at instant 5. In this case, to reach the farthest customer, the UAV flies to node 2, "takes" service 2, and then departs and flies to the customer. After delivery, it "takes" service-3, transfers to service-4 and returns to the depot. Since at the beginning the UAV has to fly to node 2, much energy has been consumed. Thus, it only increases the coverage by about 10%. In Figure 6.16d, we consider the case where service-2 breaks down at instant 10. Compared to Figure 6.16b, the extended coverage reduced from 30% to about 15%. It is clear that the break down of service-2 impacts on the coverage extension less than the break down of service-1. If the service-1 or 2 recovers later, the extended coverage will be as shown in Figure 6.16b. From this analysis we can see that traffic freeze makes some transportation vehicles unable to assist UAV delivery, but the rest of the transportation vehicles can still help to increase the coverage area.

6.6 Round-trip Path Planning in a Stochastic Network

Following the discussion on the round-trip path planning problem in a deterministic network in Section 6.5, this section discusses the more general and also more challenging round-trip path planning problem in a stochastic network.

6.6.1 Problem Statement

The energy consumption rates in these status are assumed to be constants, denoted by P_v, P_f, and P_w, respectively. For the first two status, the energy consumption can be added to the corresponding links. Let $e_{ij}(t)$ denote the energy consumption on the link a_{ij} when leaving at instant t. If a_{ij} is a normal link, $e_{ij}(t) = P_v b_{ij}(t)$; if a_{ij} is an extra link, $e_{ji}(t) = P_f b_{ij}(t)$. To model the energy consumption in the last status, some new notations are needed. Let r_u denote the instant at which the UAV reaches node u and l_u denote the instant at which the UAV leaves node u. Let $w_{ij}(t)$ denote the waiting time at node u when the UAV reaches u at instant t, and the UAV plans to traverse the link a_{ij}. Then, the energy consumption during this waiting period is given by $P_w w_{ij}(t)$. With these notations, one can have:

$$l_u = r_u + w_{ij}(r_u), \tag{6.31}$$

$$r_v = l_u + b_{ij}(l_u). \tag{6.32}$$

These relations are also indicated in Figure 6.17. Substituting Eq. (6.31) into Eq. (6.32), the following relationship can be obtained:

$$r_v = r_u + w_{ij}(r_u) + b_{ij}(r_u + w_{ij}(r_u)). \tag{6.33}$$

In other words, Eq. (6.33) gives the relationship between the reaching instants at two consecutively visited nodes. To simplify Eq. (6.33), an intermediate notation $\lambda_{ij}(t)$ is introduced to represent the total time to complete a link, which includes the waiting time and the link traversal time, see Figure 6.17:

$$\lambda_{ij}(r_u) = w_{ij}(r_u) + b_{ij}(r_u + w_{ij}(r_u)). \tag{6.34}$$

Then, Eq. (6.33) can be written as follows:

$$r_v = r_u + \lambda_{ij}(r_u), \tag{6.35}$$

Let c_u denote the cumulative energy consumption when the UAV reaches node u from the depot D. Afterwards, the UAV will consume some energy on hovering and then on traversing the following link, say a_{ij}. The cumulative energy consumption

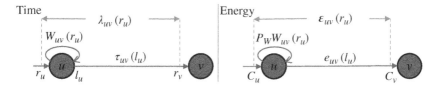

Figure 6.17 The relationship between several key notations of time instants, time periods and energy consumption.

when the UAV reaches node v is given by:

$$c_v = c_u + P_w w_{ij}(r_u) + e_{ij}(r_u + w_{ij}(r_u)). \tag{6.36}$$

Similar to the simplification on Eq. (6.33), another intermediate notation $\epsilon_{ij}(t)$ representing the total energy consumption to complete the link a_{ij} is introduced, which includes that for hovering and that for traversing the link a_{ij}:

$$\epsilon_{ij}(r_u) = P_w w_{ij}(r_u) + e_{ij}(r_u + w_{ij}(r_u)). \tag{6.37}$$

Then, Eq. (6.36) can be written as follows; see Figure 6.17:

$$c_v = c_u + \epsilon_{ij}(r_u). \tag{6.38}$$

Definition 6.5 **(Feasible path)** Let $P = \{D = u_1, \dots, u_m = C, \dots, u_n = D\}$ (where $1 < m < n$) denote a round trip path from D to C and back to D. Here, $a_{u_i u_{i+1}} \in A, \forall i = 1, 2, \dots, n - 1$. Let T_0 be a given constant signifying the customer service deadline and E_0 be a given constant representing the allowed energy consumption. If $r_{u_m = C} \leq T_0$ and $c_{u_n = D} \leq E_0$, P is a feasible path.

Remark 6.5 To avoid some trivial paths consisting of extra links only, the path $P = \{D = u_1, \dots, u_m = C, \dots, u_n = D\}$ satisfies $n \geq 5$. This requirement says that at least one normal link exists in the path. Furthermore, the paths like $P = \{D, x, C, \dots, D\}$ and $P = \{D, \dots, C, x, D\}$ are forbidden. The reason is that these paths consist of a subpath $\{D, x, C\}$ or $\{C, x, D\}$. If node x does not locate on the line connecting D and C, the subpath like $\{D, x, C\}$ definitely results in more energy consumption than the subpath $\{D, C\}$. Even if x locates on that line, it is not necessary to stop at x. Therefore, if such a subpath, like $\{D, x, C\}$ having a single node between D and C, exists on the UAV path, the node x is removed.

Problem Statement: Find the path with the minimal return instant to the depot D among all the feasible paths. In other words, for the given $G(\mathcal{N}, A)$, T_0, E_0, P_v, P_f, and P_w, look for the path $P = \{D = u_1, \cdots, u_m = C, \dots, u_n = D\}$ and the corresponding waiting time at each node and leaving instant from each node, such that the return instant to the depot is minimized while satisfying the constraints of service deadline and energy budget:

$$\min_P r_{u_n} \tag{6.39}$$

subject to

$$r_{u_m} \leq T_0, \tag{6.40}$$

$$c_{u_n} \leq E_0, \tag{6.41}$$

$$a_{u_i, u_{i+1}} \in A, \forall i = 1, \cdots, n - 1. \tag{6.42}$$

In the objective function (6.39), r_{u_n} is the return instant to D, which is computed in Eq. (6.33), so as the instant of reaching the customer r_{u_m} in constraint (6.40). Equation (6.33) is the compact form of Eqs. (6.31) and (6.32), and one can see that the reaching instant at a node r_v depends on the reaching instant at the previous node r_u, how long the UAVs waits at node u, and the time to traverse a_{ij}. Since the link traversal time $b_{ij}(t)$ is time-dependent, all the variables computed by Eqs. (6.31) and (6.32) inherit the time-dependency. Moreover, since the energy consumption relates to the link traversal time and the time for hovering, the variable c_{u_n} in Eq. (6.41) is also time-dependent. Therefore, the considered problem is time-dependent.

Remark 6.6 The formulated problem does not optimize the energy consumption, but regard it as a constraint. The reason is that many practitioners may have multiple backup batteries (as batteries are cheaper than UAVs in general) and they may prefer to recharge the battery after working hours. This can make full use of the working time in the day. The considered problem requires the UAV to return to the depot in the fast way while satisfying the constraints. The developed method will play an important role in investigating the more complex case with multiple orders and multiple UAVs. In another situation, i.e. the supplier does not have backup batteries, optimizing both time and energy consumption is crucial.

6.6.2 Proposed Solution

This section presents an exact solution algorithm to solve the problem (6.39)–(6.42). The basic idea is label setting, i.e. it keeps updating the labels of the nodes in $\mathcal{N}$ until the path is found. The algorithm dealing with an ideal case is first presented in Section 6.6.2.1. It is then extended to tackle the more practical case where the link traversal times are not only time-dependent but also uncertain in Section 6.6.2.2.

6.6.2.1 Proposed Algorithm
The fundamental idea is to update the labels associated with the nodes in the graph. There are two key techniques. The first one is a recursive formula guiding the label setting operation. The second one is an adjustment of some nodes' labels to facilitate the backward trip construction. These two techniques are presented in details below.

Definition 6.6 Let $P = \{D = u_1, \ldots, u_n = u\}$ be a path from the depot D to u. If there exists some waiting time at each node of P such that the UAV reaches node u at an instant no greater than t, i.e. $r_u \leq t$, path P is said to have time t.

Remark 6.7 At instant t, the actual instant at which the UAV reaches node u can be t or smaller than t. This happens frequently in transit networks. The actual reaching instant at a stop belongs to a set of discrete instants. For instance, two vehicles arrive at a stop at instants 5 and 10, respectively. When the reaching instant is inquired at instant 7, there is no vehicle arrived. However, with Definition 6.5, there was a service arriving at instant 5, when it is inquired at instant 7.

The notations about time periods and instants defined in Section 6.6.1 are assumed to be integers. This is easy to be satisfied since a suitable unit time can be found, say 10 seconds. Then, all the relevant notations will be some integer times of this unit. Furthermore, it is assumed that $0 < P_v < P_w < P_f$. In other words, the energy consumption rate when riding on a vehicle is less than hovering, which is further less than flying. This assumption is practical, since when the UAV is riding on a bus or a train on the roof, the UAV can simply turn off its motors. As a return, the saved energy increases the UAV's effective travel range. This is the most important advantage of the transit network based UAV delivery.

Before presenting the recursive formula, three key labels associated with each node in $\mathcal{N}$ are introduced:

- $\mathcal{T}(u, t)$: The actual instant at which the UAV reaches node u when it is inquired at instant t. According to Definition 6.5, $\mathcal{T}(u, t) \leq t$.
- $\mathcal{E}(u, t)$: The cumulative energy consumption of the UAV when it reaches node u following a path P inquired at instant t.
- $P(u, t)$: The predecessor node of u on the best path from D to u which has time t.

Recursive formula: Consider a path from D to u that has time k. This path is extended to a node v by adding a link $a_{ij} \in \mathcal{A}$, such that the extended path has time t. Let $Q_{ij}(t)$ denote the energy consumption of the expended path, which is computed as follows:

$$Q_{ij}(t) = \min_{\{k + \lambda_{ij}(k) = t\}} \{\mathcal{E}(i, k) + \epsilon_{ij}(k)\}. \tag{6.43}$$

An illustration of Eq. (6.43) is shown in Figure 6.18. From node i to node j, there may exist multiple links that can make the reaching instant at j be t. Among them, the link resulting in the lowest energy consumption should be selected.

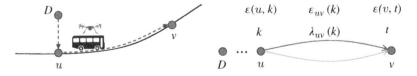

Figure 6.18 Extending the path from D to u by adding a link a_{ij}.

Introduce the following recursive relationship to update the accumulated energy consumption when the path reaches j:

$$\mathcal{E}(j,t) = \min\{\mathcal{E}(j,t-1),$$
$$\min_{\{i|a_{ij}\in\mathcal{A}\}}\min_{\{k|k+\lambda_{ij}(k)=t\}}\{\mathcal{E}(i,k)+\epsilon_{ij}(k)\}\}. \tag{6.44}$$

With Eq. (6.43), Eq. (6.44) can be simplified as follows:

$$\mathcal{E}(j,t) = \min\{\mathcal{E}(v,t-1), \min_{\{i|a_{ij}\in\mathcal{A}\}}\mathcal{Q}_{ij}(t)\}. \tag{6.45}$$

The intuition of Eqs. (6.44) or (6.45) is twofold. Firstly, if there is not a path that can reach node j and have time t, $\mathcal{E}(j,t)$ is updated by the previously recorded value, i.e. $\mathcal{E}(j,t) = \mathcal{E}(j,t-1)$, which is the first part of Eq. (6.45). Here, the path reaching node j has time $t-1$, i.e. $r_v \leq t-1$. Of course, such a path also has time t, since $r_v \leq t-1 < t$. Secondly, if the following conditions hold: (i) there exists a path from D to node i, and the path has time k, (ii) $a_{ij} \in \mathcal{A}$, and (iii) there further exists a waiting time and a link traversal time such that $k + \lambda_{ij}(k) = t$, this path can be extended by adding the waiting time to node i and a link a_{ij} to the end of the path. Then, the path reaches node j at time t, and the cumulative energy consumption on the extended path is given by $\mathcal{E}(i,k) + \epsilon_{ij}(k)$. If there are multiple combinations of j, the waiting time, and the link traversal time, the one corresponding to the smallest energy consumption should be taken, which is the second part of Eq. (6.45). Therefore, $\mathcal{E}(j,t)$ should take the smaller one of the above two cases. An illustrative example of updating the label of $\mathcal{E}(j,t)$ is presented in Figure 6.19. On the left hand side, the computed $\mathcal{Q}_{ij} = 18$, which is smaller than $\mathcal{E}(j,11)$. Thus, according to Eq. (6.45), $\mathcal{E}(j,12)$ takes the label of $\mathcal{Q}_{ij}$, i.e. 18. This means that a better path is found to reach node j via node u, and the reaching instant is 12. On the right hand side, $\mathcal{Q}_{ij} = 21$, which is larger than $\mathcal{E}(j,11)$. Thus, $\mathcal{E}(j,12)$ keeps the recording of $\mathcal{E}(j,11)$, i.e. 20. This means that no better path is found to reach node j at time 12.

To compute the intermediate notations $\lambda_{ij}(k)$ and $\epsilon_{ij}(k)$ in (6.44), the actual reaching instant is required, which is stored in $\mathcal{T}(i,t)$. In general, if there is a

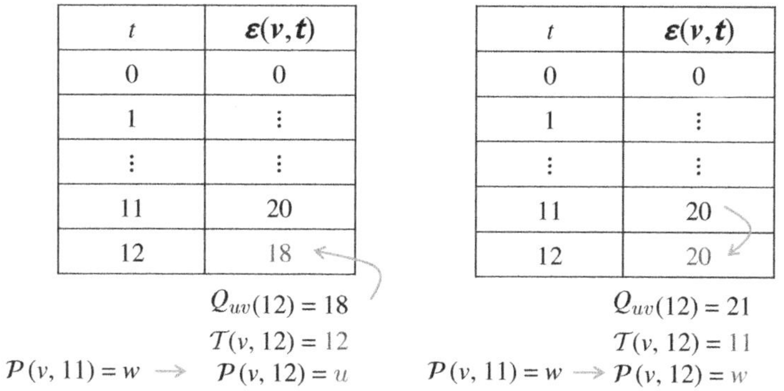

t	$\varepsilon(v,t)$
0	0
1	$\vdots$
$\vdots$	$\vdots$
11	20
12	18

$\mathcal{Q}_{uv}(12) = 18$
$\mathcal{T}(v,12) = 12$
$\mathcal{P}(v,11) = w \;\rightarrow\; \mathcal{P}(v,12) = u$

t	$\varepsilon(v,t)$
0	0
1	$\vdots$
$\vdots$	$\vdots$
11	20
12	20

$\mathcal{Q}_{uv}(12) = 21$
$\mathcal{T}(v,12) = 11$
$\mathcal{P}(v,11) = w \;\rightarrow\; \mathcal{P}(v,12) = w$

Figure 6.19 An example of updating $\mathcal{E}(j,t)$, $\mathcal{T}(j,t)$ and $\mathcal{P}(j,t)$.

path from D to i that has time t, then $\mathcal{T}(i, t) = t$; otherwise $\mathcal{T}(i, t)$ stores the actual arrival instant at node i before t, and such a time equals to the value stored at instant $t - 1$, i.e, $\mathcal{T}(i, t) = \mathcal{T}(i, t - 1)$. With this actual reaching instant, the waiting time at node i can be further computed and so as the corresponding energy consumption for waiting. Specifically, the waiting time at node i is given by $t - \mathcal{T}(i, t)$ for the currently considered instant t and the corresponding energy consumption for waiting equals $P_w(t - \mathcal{T}(i, t))$. Along with updating $\mathcal{E}(j, t)$ by Eq. (6.45), $\mathcal{P}(j, t)$ is updated. Specifically, $\mathcal{P}(j, t)$ records the node that makes the second part of Eq. (6.45) the smallest or the node stored in $\mathcal{P}(j, t - 1)$. Figure 6.19 also illustrates the updates of $\mathcal{T}(j, t)$ and $\mathcal{P}(j, t)$. For the instant 12, when $\mathcal{E}(j, 12)$ is updated by $Q_{ij}(12)$, a better path is found to reach node j at instant 12. Thus, $\mathcal{T}(j, 12)$ records the instant 12. Suppose a path has already been found to reach node j at instant 11 and this path is via node D. Then, $\mathcal{P}(j, 12)$ should replace D by u. In the case shown on the right hand side, the label $\mathcal{E}(j, 12)$ is not updated by $Q_{ij}(12)$. So, the labels of $\mathcal{T}(j, t)$ and $\mathcal{P}(j, t)$ both remain their previous recordings.

Label Adjustment The key operation of updating the nodes' labels has been presented. However, this operation cannot distinguish the forward trip and backward trip. Suppose that $P = \{D, \ldots, j, C, j, \ldots, D\}$ is the best path for the UAV to serve customer C. In the forward trip, the UAV reaches C from node j. After dropping off the parcel, the UAV goes to node j. However, the label of $\mathcal{E}(j, t)$ cannot be updated correctly by the recursive formula (6.45). The reason is clear: since traveling from node j to C and back to j must consume energy, the label $\mathcal{E}(j, t)$ will never be updated by the second term of Eq. (6.45). A special operation is conducted at a particular instant to address this issue.

The particular instant refers to the instant t at which the UAV reaches C following a path. Clearly, $t \leq T_0$ holds, if there exists a feasible path. From $t + 1$, one can consider to construct the backward path. The special operation refers to temperately replacing the stored labels of $\mathcal{E}(i, t)$, $\mathcal{T}(i, t)$ and $\mathcal{P}(i, t)$ for any node $i \neq C$ by E_0, ∞, and $\emptyset$, respectively. After executing this special operation at that particular instant, $\mathcal{E}(i, t + 1)$ will be updated by the second part of Eq. (6.45).

Algorithm Presentation Now, there are enough notations to present the proposed algorithm. It consists of two phases: an initial phase and a main phase. Line 1–5 is the initial phase, which sets the initial labels of each node in $\mathcal{N}$. At instant 0, since it is impossible to reach any node other than D, the cumulative energy consumption for these nodes except D is initialized as E_0 while that for D is 0. The reaching instants for all the nodes instead of D are set as infinity, and that for D is zero. $\mathcal{P}(i, 0)$ stores the empty set $\emptyset$ for any $i \neq D$ and $\mathcal{P}(D, 0)$ stores D. The *flag*(i) in Line 4 will be used to detect the particular instant.

The main phase, i.e. Line 6–24, has three parts. In the first for-loop, for each link $a_{ij} \in \mathcal{A}$, it sets $Q_{ij}(t)$ as ∞ as an initial value for the currently considered instant t.

In the second for-loop, the algorithm updates $Q_{ij}(t)$ according to Eq. (6.43), based on the previously stored $\mathcal{E}(i, k)$ and $\mathcal{T}(i, k)$, where $k < t$. In the third for-loop, the algorithm updates $\mathcal{E}(i, t)$ according to Eq. (6.45), based on the updated $Q_{ij}(t)$. $\mathcal{T}(i, t)$ stores the currently considered instant t if $\mathcal{E}(i, t)$ is updated with the second part in Eq. (6.45) (see Line 21); otherwise, $\mathcal{T}(i, t)$ stores the value of $\mathcal{T}(i, t - 1)$ (see Line 19). Notice that the first operation in Line 21 returns the node that minimizes $Q_{ij}(t)$, while the third operation returns the corresponding energy consumption value. The special operation is executed in Line 15-17. At instant t, if the condition $\mathcal{T}(j, t - 1) \leq T_0$ holds, it means that a path has been found which can reach the customer C before the deadline T_0. Then, $\mathcal{E}(j, t - 1)$, $\mathcal{T}(j, t - 1)$, and $\mathcal{P}(j, t - 1)$ for any node $i \neq C$ are replaced by E_0, ∞, and $\emptyset$, respectively, see Line 18, so that the values for instant t can be updated correctly for the backward path. Then, the notation $flag(j)$ is replaced by 0. After this particular instant, the special operation is skipped since $flag(j) == 1$ in Line 15 will not hold again.

Algorithm 6.4 The proposed algorithm.

1: $t = 0$.
2: **for** any node $i \in \mathcal{N}$ **do**
3:     $\mathcal{E}(i, t) = E_0, \mathcal{E}(D, t) = 0, \mathcal{T}(i, t) = \infty$
4:     $\mathcal{T}(D, t) = t, \mathcal{P}(i, t) = \emptyset, \mathcal{T}(D, t) = D, flag(i) = 1$.
5: **end for**
6: **while** $(\mathcal{T}(C, t) > T_0 | \mathcal{T}(C, t) \leq T_0 \,\&\, \mathcal{T}(D, t) > \bar{T}) \,\&\, (\min_u \mathcal{E}(i, t) \leq E_0)$ **do**
7:     $t = t + 1$.
8:     **for** any $a_{ij} \in \mathcal{A}$ **do**
9:         $Q_{ij}(t) = \infty$.
10:     **end for**
11:     **for** any $a_{ij} \in \mathcal{A}$ and any k such that $\mathcal{T}(i, k) + \lambda_{ij}(\mathcal{T}(i, k)) = t$ **do**
12:         $Q_{ij}(t) = \min\{Q_{ij}(t), \mathcal{E}(i, \mathcal{T}(i, k)) + \epsilon_{ij}(\mathcal{T}(i, k))\}$.
13:     **end for**
14:     **for** any $j \in \mathcal{N}$ **do**
15:         **if** $\mathcal{T}(C, t - 1) \leq T_0 \,\&\, j \neq C \,\&\, flag(j) == 1$ **then**
16:             $\mathcal{E}(j, t - 1) = E_0; \mathcal{T}(j, t - 1) = \infty; \mathcal{P}(j, t - 1) = \emptyset; flag(j) = 0$.
17:         **end if**
18:         **if** $\mathcal{E}(j, t - 1) < \min_{a_{ij} \in \mathcal{A}} Q_{ij}(t)$ **then**
19:             $\mathcal{E}(j, t) = \mathcal{E}(j, t - 1); \mathcal{T}(j, t) = \mathcal{T}(j, t - 1); \mathcal{P}(j, t) = \mathcal{P}(j, t - 1)$.
20:         **else**
21:             $\mathcal{E}(j, t) = \min_{a_{ij} \in \mathcal{A}} Q_{ij}(t); \mathcal{T}(j, t) = t; \mathcal{P}(j, t) = \min_{\{i | a_{ij} \in \mathcal{A}\}} Q_{ij}(t)$.
22:         **end if**
23:     **end for**
24: **end while**

The main phase continues if all the following two conditions are satisfied, see Line 6. The first one is that the customer C has not been reached, i.e. $\mathcal{T}(C, t) > T_0$, or the customer C has been reached but the depot D has not been reached from the backward trip, i.e. $\mathcal{T}(C, t) \leq T_0$ and $\mathcal{T}(D, t) > \overline{T}$, where $\overline{T} = \left\lceil \frac{E_0}{P_j} \right\rceil$. The explanation is as follows. Suppose the earliest return instant to the depot D is t_*. The maximum of t_* is achieved when the battery of the UAV is used for the longest time. As $P_j < P_f$, the longest time is achieved when the UAV rides on vehicles all the time and the corresponding time is given by $\overline{T} = \left\lceil \frac{E_0}{P_j} \right\rceil$. Then, $t_* \leq \overline{T}$. The second condition is that the cumulative energy consumption to at least one node is below E_0, i.e. $\min_u \mathcal{E}(i, t) < E_0$. If the cumulative energy consumption to any node is greater than E_0, it means that there does not exist a feasible path.

When the algorithm terminates with the first condition, the earliest return instant to D on the backward trip is recorded in $\mathcal{T}(D, t)$, and the corresponding cumulative energy consumption is stored in $\mathcal{E}(D, t)$. To further construct the optimal path and obtain the leaving instant from each node and the waiting time, a standard backtracking operation can be used. To support the backtracking operation, the stored values in $\mathcal{E}(i, t)$, $\mathcal{T}(i, t)$, and $\mathcal{P}(i, t)$ will be used.

Computational Complexity In the initial phase, as there are totally $O(|\mathcal{N}|)$ nodes to initialize, this phase can be done in $O(|\mathcal{N}|)$ time, where $|\mathcal{N}|$ is the number of nodes in $\mathcal{N}$. The main phase will be repeated for up to $O(\overline{T})$ times. In the first part of the main phase, for any t, this part can be done in $O(|\mathcal{A}|)$, where $|\mathcal{A}|$ is the number of links in $\mathcal{A}$. Then, the complexity is $O(\overline{T}|\mathcal{A}|)$. In the second part, for any t, the algorithm accounts all the links in $\mathcal{A}$ and all the instants k that make $\mathcal{T}(i, k) + \lambda(i, j, \mathcal{T}(i, k)) = t$ hold. Thus, there exist up to $O(\overline{T}|\mathcal{A}|)$ combinations of a link and an instant. Since the intermediate parameters (including $\mathcal{T}(i, k)$ and $\mathcal{E}(i, k)$ where $k < t$) have already been computed before, updating $Q_{ij}(t)$ can be done in $O(1)$. Then, the complexity of the second part is $O(\overline{T}|\mathcal{A}|)$. The third part executes the special operation, and updates $\mathcal{T}(j, t)$, $\mathcal{E}(j, t)$, and $\mathcal{P}(j, t)$ for any node j except C and each update can be done in $O(1)$. Thus, the complexity is $O(\overline{T}|\mathcal{N}|)$. Therefore, the overall complexity is $O(|\mathcal{N}| + 2\overline{T}|\mathcal{A}| + \overline{T}|\mathcal{N}|) = O(\overline{T}(|\mathcal{N}| + |\mathcal{A}|))$.

6.6.2.2 Robust Round-trip Planning Algorithm
The algorithm presented above focuses on the situation where the information in $\mathcal{G}(\mathcal{N}, \mathcal{A})$ is actually known. However, the reality is that the vehicles cannot exactly follow their timetables in general. Thus, it is necessary to take the randomness into account to design a round trip UAV path which is robust to the uncertain transit network.

The robust version of the round-trip planning algorithm requires the statistic data of the transit network and the UAV flights to build up the distribution

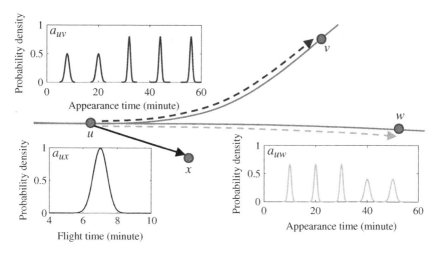

Figure 6.20 Vehicle appearance time distributions and distribution of flight time. a_{ij}, a_{iw}: normal links; a_{ix}: extra link.

functions of the vehicle arrival and departure times and the UAV flying times. Regarding the statistic data of the transit network, it can be obtained via the onboard GPS module, which can record the arrival time at a stop and the departure time. Then, each stop node is not only associated with some time instants (in the timetable) but also their distributions for the connected normal links; see Figure 6.20. For the UAV flying times, field experiments should be carried out in different weather conditions to find the distributions of the flying times; see Figure 6.20.

In the public transportation vehicles supported delivery mode, flying alone and riding on a vehicle alternate on a path. When the link traversal times are uncertain, a practical condition to construct a reliable path is: the probability that the UAV arrival time at a stop (by flying) is earlier than the vehicle departure time should be as large as possible. To make it convenient to verify this reliability, two more pieces of information are added to each node $v \in \mathcal{N}$, i.e. two "end" instants of the appearance time. Let $\rho_0 \in (0, 1)$ be a given threshold. The two "end" instants t_1 and t_2 are defined as follows. t_1 is one "end" instant such that the probability that a vehicle appears at a node before t_1 is no larger than $1 - \rho_0$, and t_2 is the other "end" instant such that the probability that a vehicle appears at a node after t_2 is no larger than ρ_0. t_1 represents the "earliest" arrival instant, and t_2 represents the "latest" departure instant. This definition also applies to the arrival instant of the UAV path: t_1 and t_2 indicate the "earliest" and "latest" arrival instants, respectively; see Figure 6.21.

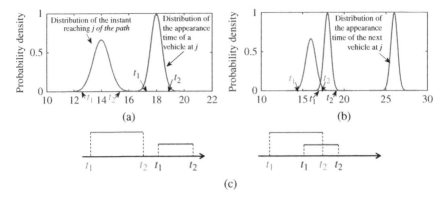

Figure 6.21 The robust version to update labels of nodes. (a) Node j can be updated. (b) Node j cannot be updated with the information relating to the first vehicle. (c) Another representation of whether j's labels can be updated.

In the robust version, when node j's labels are to be updated, the two "end" instants are used. Specifically, if the latest arrival instant of the path at j is earlier than the earliest arrival instant of a vehicle at j, the labels of j can be updated; see in Figure 6.21a. Otherwise, the labels of j have to be updated by the information of the next vehicle; see Figure 6.21b. Note that it is possible for the UAV to ride on the first vehicle in the case of Figure 6.21b, since the "latest" arrival instant is early than the "latest" departure instant. However, it is also possible for the UAV to miss the vehicle at node j, see Figure 6.21c.

The framework of Algorithm 6.4 can be extended to the robust case with the following modifications. First, the path arrival time is given by the "latest" arrival instant. Second, any vehicle departure time is given by the "earliest" arrival instant. Third, the accumulative energy consumption of the path is modified according to the "latest" arrival instant. Then, the generated path by the modified Algorithm 6.4 is robust in the sense of ρ_0. It is clear that the larger the ρ_0, the more robust the path.

Remark 6.8 The uncertain and predictable case of the transit network has been considered. In practice, many unpredictable traffic accidents occur from time to time. In this situation, some part of the transit network freezes. If the UAV uses the affected part of the network and it is already on the way, it may not be able to reach the customer or return to the depot, since the residual energy does not allow it to do so by flying only. A contingency plan is to fly to the roof of a nearby building and wait for collection by human operators, which guarantees the safety of the public and the UAV itself, since the roof is generally unused. As a result, the customer may not receive the order on time. The supplier needs to inform the customer and use other methods such as the conventional ground vehicle delivery. This issue is not unique to UAV delivery. A ground delivery vehicle has a higher probability to suffer from traffic conditions than the considered UAV delivery mode, since only a part of the UAV path relies on the transit network while the whole path of a ground vehicle depends on the road conditions.

6.6.3 Evaluation

In this section, how the proposed method works is demonstrated first. In the toy example shown in Figure 6.22a, there are three stops and six routes. Figure 6.22b covers three typical cases for the link traversal time $b_{ij}(t)$. The traversal times of a_{xy} and a_{yx} keep constant, those of a_{xz} and a_{zx} increase with t, and those of a_{zy} and a_{yz} decrease with t. Differently, the extra links are with fixed traversal times, see Figure 6.22a. The service deadline T_0 is set as 12 minutes. The energy consumption rates are set as $P_v = 1$, $P_w = 4$, and $P_f = 5$ unit per minute, and $E_0 = 50$. Clearly, the customer C cannot be reached by flying only, since the UAV has to fly for more than 14 minutes and the required energy consumption is over E_0.

Algorithm 6.4 is applied, and the nodes' labels are shown in Table 6.5. The UAV reaches C in 10 minutes and returns to D in 19 minutes. The total energy consumption is 41 units. To find the path, the backward tracking method is used, and the nodes' labels are considered together with the link traversal times. For example, the last node on the path is D, and its predecessor node is x; see the last line of Table 6.5. From Figure 6.22a, one can see that from x to D, it takes one minute and the corresponding energy consumption is 5 units for flying. Hence, the UAV departs node x at instant 18 (19-1), and at this instant, the accumulative energy

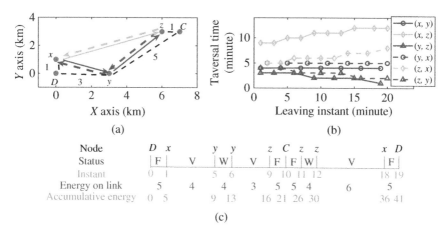

Node	D	x		y	y		z	C	z	z		x	D
Status	F		V	W		V	F	F	W		V		F
Instant	0	1		5	6		9	10	11	12			18	19
Energy on link	5		4		4		3	5	5	4		6		5
Accumulative energy	0	5		9	13		16	21	26	30			36	41

(c)

Figure 6.22 (a) The example used for evaluation. (b) The traversal times of the normal links. (c) The UAV path.

Table 6.5 Stored values of $\mathcal{E}(u,t)$, $\mathcal{T}(u,t)$ and $\mathcal{P}(u,t)$.

t	$\mathcal{E}(D,t)/$ $\mathcal{T}(D,t)/$ $\mathcal{P}(D,t)$	$\mathcal{E}(x,t)/$ $\mathcal{T}(x,t)/$ $\mathcal{P}(x,t)$	$\mathcal{E}(y,t)/$ $\mathcal{T}(y,t)/$ $\mathcal{P}(y,t)$	$\mathcal{E}(z,t)/$ $\mathcal{T}(z,t)/$ $\mathcal{P}(z,t)$	$\mathcal{E}(C,t)/$ $\mathcal{T}(C,t)/$ $\mathcal{P}(C,t)$
0	0/ 0/ D	∞/ ∞/ ∅	∞/ ∞/ ∅	∞/ ∞/ ∅	∞/ ∞/ ∅
1	0/ 1/ D	5/ 1/ D	∞/ ∞/ ∅	∞/ ∞/ ∅	∞/ ∞/ ∅
2	0/ 2/ D	5/ 2/ D	∞/ ∞/ ∅	∞/ ∞/ ∅	∞/ ∞/ ∅
3	0/ 3/ D	5/ 3/ D	15/ 3/ D	∞/ ∞/ ∅	∞/ ∞/ ∅
4	0/ 4/ D	5/ 4/ D	15/ 4/ D	∞/ ∞/ ∅	∞/ ∞/ ∅
5	0/ 5/ D	5/ 5/ D	9/ 5/ x	∞/ ∞/ ∅	∞/ ∞/ ∅
6	0/ 6/ D	5/ 6/ D	9/ 5/ x	∞/ ∞/ ∅	∞/ ∞/ ∅
7	0/ 7/ D	5/ 7/ D	9/ 5/ x	∞/ ∞/ ∅	∞/ ∞/ ∅
8	0/ 8/ D	5/ 8/ D	9/ 8/ x	∞/ ∞/ ∅	∞/ ∞/ ∅
9	0/ 9/ D	5/ 9/ D	9/ 8/ x	16/ 9/ y	∞/ ∞/ ∅
10	50 (0)/ ∞ (10)/ ∅ (D)	50 (5)/ ∞ (10)/ ∅ (D)	50 (9)/ ∞ (8)/ ∞ (10)	50 (16)/ ∞ (9)/ ∅ (y)	**21/ 10/ z**
11	∞/ ∞/ ∅	∞/ ∞/ ∅	∞/ ∞/ ∅	26/ 11/ C	21/ 10/ z
12	∞/ ∞/ ∅	∞/ ∞/ ∅	∞/ ∞/ ∅	26/ 11/ C	21/ 10/ z
13	∞/ ∞/ ∅	∞/ ∞/ ∅	∞/ ∞/ ∅	26/ 11/ C	21/ 10/ z
14	∞/ ∞/ ∅	∞/ ∞/ ∅	∞/ ∞/ ∅	26/ 11/ C	21/ 10/ z
15	∞/ ∞/ ∅	∞/ ∞/ ∅	46/ 15/ C	26/ 11/ C	21/ 10/ z
16	∞/ ∞/ ∅	∞/ ∞/ ∅	46/ 15/ C	26/ 11/ C	21/ 10/ z
17	∞/ ∞/ ∅	∞/ ∞/ ∅	46/ 15/ C	26/ 11/ C	21/ 10/ z
18	∞/ ∞/ ∅	36/ 18/ z	46/ 15/ C	26/ 11/ C	21/ 10/ z
19	**41/ 19/ x**	36/ 18/ z	46/ 15/ C	26/ 11/ C	21/ 10/ z

consumption is 36 (41-5). With this backward tracking method, the whole path can be found as shown in Figure 6.22c.

Moreover, the robust version of Algorithm 6.4 is tested in a real-world case with the public transportation network in Sydney. Suppose a depot locates at the University of New South Wales (UNSW), and a customer locates at Westmead, which is about 22 km away from UNSW (straight line distance on Google Maps); see Figure 6.23a. Clearly, this customer cannot be served by a typical Amazon UAV by flying only. There are various public transportation services passing the depot and the customer. To form a connected network, they are linked to the nearby stops, and some stops are also linked. Since the historical data of the public

Time (in minute): 80
Energy consumption: 58.17

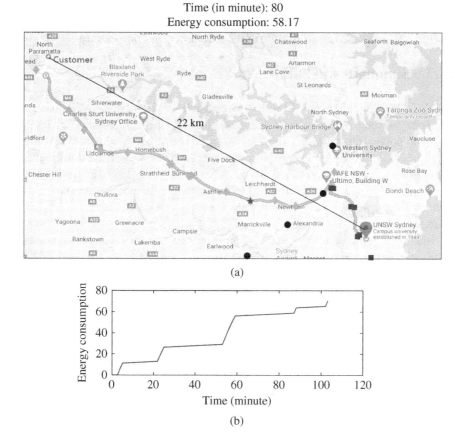

(a)

(b)

Figure 6.23 (a) Round trip path in the real-world transit network. The UAV is marked by the pentagon. (b) Energy consumption.

transportation vehicles is not available, the appearance time of a vehicle at a stop is assumed to be one minute around the scheduled instant for trains and two minutes for buses, since trains are often more reliable than buses. Then, from the scheduled instant (which is available from the public timetable), the "earliest" arrival instant and "latest" departure instant defined in Section 6.6.2.2 are known. Applying the robust version of Algorithm 6.4, the UAV path for the considered customer is found. The simulated movement of the UAV from the depot to the customer and then back to the depot is shown in a video (https://youtu.be/dSdXcPSNTF0). It takes the UAV about 100 minutes to complete the delivery. The energy consumption during this period is shown in Figure 6.23b. For the most time, the energy consumption is low, because the UAV is riding on a vehicle at that time. The UAV only flies and hovers for about 14 minutes during this mission, which is achievable by a commercial delivery UAV. The saved energy by the vehicles' transportation further increases the tolerant ability of the transit network uncertainty.

6.7 Summary and Future Work

In this chapter, we presented the recent research work on using UAVs to conduct the last-mile delivery. In particular, we focused on the integration of UAVs and the public transportation vehicles. We presented three methods. While the first method targets the one-way path planning, the second and the third methods targets the round-trip path planning. A particular focus has been paid to the time-dependency and schematics of the timetables of the public transportation vehicles. The effectiveness of the discussed methods have been verified via computer simulations.

The main limitation of the discussed approaches is that they only provide solutions to delivering one order by one UAV. The obtained results do not address the complex case with multiple orders and multiple drones. Since the public transportation network is time-dependent, the delivery time and the energy consumption are also time-dependent, so as the high-level scheduling problem. Thus, the future research will be on the time-dependent scheduling problem with multiple orders, depots, and drones. This is to address the allocation of orders and the determination of the delivery sequence of the allocated orders.

References

1 Statista, "Retail e-commerce sales worldwide from 2014 to 2021," Accessed on 1 June 2021. Online: https://www.statista.com/statistics/379046/worldwide-retail-e-commerce-sales/.

2 J. Allen, M. Piecyk, M. Piotrowska, F. McLeod, T. Cherrett, K. Ghali, T. Nguyen, T. Bektas, O. Bates, A. Friday, *et al.*, "Understanding the impact of e-commerce on last-mile light goods vehicle activity in urban areas: the case of London," *Transportation Research Part D: Transport and Environment*, vol. 61, pp. 325–338, 2018.

3 H. Huang and A. V. Savkin, "Towards the internet of flying robots: a survey," *Sensors*, vol. 18, no. 11, p. 4038, 2018.

4 Amazon.com Inc., "Amazon prime air," Accessed on 1 Feb. 2021. Online: http://www.amazon.com/primeair.

5 "China is on the fast track to drone deliveries," Accessed on 26 June 2021. Online: https://www.bloomberg.com/news/features/2018-07-03/china-s-on-the-fast-track-to-making-uav-drone-deliveries.

6 "SF express approved to fly drones to deliver goods," Accessed on 26 June 2021. Online: https://www.caixinglobal.com/2018-03-28/sf-express-approved-to-fly-drones-to-deliver-goods-101227325.html.

7 K.-C. Ying, C.-C. Lu, and J.-C. Chen, "Exact algorithms for single-machine scheduling problems with a variable maintenance," *Computers & Industrial Engineering*, vol. 98, pp. 427–433, 2016.

8 T. Bulhoes, R. Sadykov, A. Subramanian, and E. Uchoa, "On the exact solution of a large class of parallel machine scheduling problems," *Journal of Scheduling*, vol. 23 pp. 411–429, 2020.

9 C. C. Murray and A. G. Chu, "The flying sidekick traveling salesman problem: optimization of drone-assisted parcel delivery," *Transportation Research Part C: Emerging Technologies*, vol. 54, pp. 86–109, 2015.

10 N. Mathew, S. L. Smith, and S. L. Waslander, "Planning paths for package delivery in heterogeneous multirobot teams," *IEEE Transactions on Automation Science and Engineering*, vol. 12, pp. 1298–1308, 2015.

11 K. Dorling, J. Heinrichs, G. G. Messier, and S. Magierowski, "Vehicle routing problems for drone delivery," *IEEE Transactions on Systems, Man, and Cybernetics: Systems*, vol. 47, no. 1, pp. 70–85, 2017.

12 N. Boysen, S. Schwerdfeger, and F. Weidinger, "Scheduling last-mile deliveries with truck-based autonomous robots," *European Journal of Operational Research*, vol. 271, no. 3, pp. 1085–1099, 2018.

13 N. Agatz, P. Bouman, and M. Schmidt, "Optimization approaches for the traveling salesman problem with drone," *Transportation Science*, vol. 52, no. 4, pp. 965–981, 2018.

14 S. Kim and I. Moon, "Traveling salesman problem with a drone station," *IEEE Transactions on Systems, Man, and Cybernetics: Systems*, vol. 49, no. 1, pp. 42–52, 2019.

15 Q. M. Ha, Y. Deville, Q. D. Pham, and M. H. Hà, "On the min-cost traveling salesman problem with drone," *Transportation Research Part C: Emerging Technologies*, vol. 86, pp. 597–621, 2018.

16 W.-C. Chiang, Y. Li, J. Shang, and T. L. Urban, "Impact of drone delivery on sustainability and cost: realizing the UAV potential through vehicle routing optimization," *Applied Energy*, vol. 242, pp. 1164–1175, 2019.

17 M. W. Ulmer and B. W. Thomas, "Same-day delivery with heterogeneous fleets of drones and vehicles," *Networks*, vol. 72, no. 4, pp. 475–505, 2018.

18 "Mercedes-benz vans," Accessed on 26 June 2019. Online: https://www .mercedes-benz.com/en/mercedes-benz/vehicles/transporter/vans-drones-in-zurich/.

19 Y. Wang, F. Xu, S. Mao, S. Yang, and Y. Shen, "Adaptive online power management for more electric aircraft with hybrid energy storage systems," *IEEE*

Transactions on Transportation Electrification, vol. 6, no. 4, pp. 1780–1790, 2020.

20 Z. Pan, L. An, and C. Wen, "Recent advances in fuel cells based propulsion systems for unmanned aerial vehicles," *Applied Energy*, vol. 240, pp. 473–485, 2019.

21 H. Huang and A. V. Savkin, "A method of optimized deployment of charging stations for drone delivery," *IEEE Transactions on Transportation Electrification*, vol. 6, no. 2, pp. 510–518, 2020.

22 I. Hong, M. Kuby, and A. T. Murray, "A range-restricted recharging station coverage model for drone delivery service planning,"*Transportation Research Part C: Emerging Technologies*, vol. 90, pp. 198–212, 2018.

23 H. D. Yoo and S. M. Chankov, "Drone-delivery using autonomous mobility: an innovative approach to future last-mile delivery problems," in *IEEE International Conference on Industrial Engineering and Engineering Management (IEEM)*, pp. 1216–1220, Dec 2018.

24 R. K. Ahuja, T. L. Magnanti, and J. B. Orlin, *Network flows: theory, algorithms, and applications*. Upper Saddle River, NJ: Prentice Hall, 1993.

25 Y. Zheng, Y. Zhang, and L. Li, "Reliable path planning for bus networks considering travel time uncertainty," *IEEE Intelligent Transportation Systems Magazine*, vol. 8, no. 1, pp. 35–50, 2016.

26 Y. M. Nie and X. Wu, "Shortest path problem considering on-time arrival probability," *Transportation Research Part B: Methodological*, vol. 43, no. 6, pp. 597–613, 2009.

27 H. Huang, A. V. Savkin, and C. Huang, "Reliable path planning for drone delivery using a stochastic time-dependent public transportation network," *IEEE Transactions on Intelligent Transportation Systems*, vol. 22, no. 8, pp. 4941–4950, 2021.

28 H. Huang, A. V. Savkin, and C. Huang, "Round trip routing for energy-efficient drone delivery based on a public transportation network," *IEEE Transactions on Transportation Electrification*, vol. 6, no. 3, pp. 1368–1376, 2020.

29 H. Huang, A. V. Savkin, and C. Huang, "Drone routing in a time-dependent network: toward low-cost and large-range parcel delivery," *IEEE Transactions on Industrial Informatics*, vol. 17, no. 2, pp. 1526–1534, 2021.

30 F. Glover, G. Gutin, A. Yeo, and A. Zverovich, "Construction heuristics for the asymmetric TSP,"*European Journal of Operational Research*, vol. 129, no. 3, pp. 555–568, 2001.

31 T. Zhang, L. Ke, J. Li, J. Li, J. Huang, and Z. Li, "Metaheuristics for the tabu clustered traveling salesman problem," *Computers & Operations Research*, vol. 89, pp. 1–12, 2018.

32 J. F. Ehmke, A. M. Campbell, and T. L. Urban, "Ensuring service levels in routing problems with time windows and stochastic travel times," *European Journal of Operational Research*, vol. 240, no. 2, pp. 539–550, 2015.

33 K. Braekers, K. Ramaekers, and I. Van Nieuwenhuyse, "The vehicle routing problem: state of the art classification and review," *Computers & Industrial Engineering*, vol. 99, pp. 300–313, 2016.

34 S. M. Shavarani, M. G. Nejad, F. Rismanchian, and G. Izbirak, "Application of hierarchical facility location problem for optimization of a drone delivery system: a case study of Amazon prime air in the City of San Francisco," *International Journal of Advanced Manufacturing Technology*, vol. 95, no. 9–12, pp. 3141–3153, 2018.

35 J. Shao, J. Cheng, B. Xia, K. Yang, and H. Wei, "A novel service system for long-distance drone delivery using the "Ant Colony+A*" algorithm," *IEEE Systems Journal*, vol. 15, no. 3, pp. 3348–3359, 2021.

36 D. Sun, X. Peng, R. Qiu, and Y. Huang, "The traveling salesman problem: route planning of recharging station-assisted drone delivery," in *Proceedings of the 14th International Conference on Management Science and Engineering Management*, pp. 13–23, Springer International Publishing, 2021.

37 H. Huang, A. V. Savkin, and C. Huang, "A new parcel delivery system with drones and a public train," *Journal of Intelligent and Robotic Systems*, vol. 100, no. 3, pp. 1341–1354, 2020.

38 A. R. Al-Ali, I. Zualkernan, and F. Aloul, "A mobile GPRS-sensors array for air pollution monitoring," *IEEE Sensors Journal*, vol. 10, no. 10, pp. 1666–1671, 2010.

39 H. Huang, A. V. Savkin, and C. Huang, "I-UMDPC: the improved-unusual message delivery path construction for wireless sensor networks with mobile sinks," *IEEE Internet of Things Journal*, vol. 4, pp. 1528–1536, 2017.

40 A. Trotta, F. D. Andreagiovanni, M. Di Felice, E. Natalizio, and K. R. Chowdhury, "When UAVs ride a bus: towards energy-efficient city-scale video surveillance," in *IEEE Conference on Computer Communications*, pp. 1043–1051, IEEE, 2018.

41 H. Huang, A. V. Savkin, and C. Huang, "Scheduling of a parcel delivery system consisting of an aerial drone interacting with public transportation vehicles," *Sensors*, vol. 20, no. 7, p. 2045, 2020.

42 P. Yang, X. Cao, C. Yin, Z. Xiao, X. Xi, and D. Wu, "Proactive drone-cell deployment: overload relief for a cellular network under flash crowd traffic," *IEEE Transactions on Intelligent Transportation Systems*, vol. 18, pp. 2877–2892, 2017.

43 H. Huang and A. V. Savkin, "An algorithm of reactive collision free 3-D deployment of networked unmanned aerial vehicles for surveillance and monitoring," *IEEE Transactions on Industrial Informatics*, vol. 16, no. 1, pp. 132–140, 2020.

44 "UPS testing drones for use in its package delivery system," Accessed on 26 June 2021. Online: https://www.apnews.com/f34dc40191534203aa5d041c 3010f6c5.

45 "DHL's parcelcopter: changing shipping forever," Accessed on 26 June 2021. Online: https://discover.dhl.com/business/business-ethics/parcelcopter-drone-technology.

46 M. Torabbeigi, G. J. Lim, and S. J. Kim, "Drone delivery schedule optimization considering the reliability of drones," in *2018 International Conference on Unmanned Aircraft Systems (ICUAS)*, pp. 1048–1053, IEEE, 2018.

47 P. Grippa, D. A. Behrens, F. Wall, and C. Bettstetter, "Drone delivery systems: job assignment and dimensioning," *Autonomous Robots*, vol. 43, no. 2, pp. 261–274, 2019.

48 Z. Wang and J.-B. Sheu, "Vehicle routing problem with drones," *Transportation Research Part B: Methodological*, vol. 122, pp. 350–364, 2019.

49 S. M. Shavarani, S. Mosallaeipour, M. Golabi, and G. İzbirak, "A congested capacitated multi-level fuzzy facility location problem: an efficient drone delivery system," *Computers & Operations Research*, vol. 108, pp. 57–68, 2019.

50 E. W. Dijkstra, "A note on two problems in connexion with graphs," *Numerische Mathematik*, vol. 1, no. 1, pp. 269–271, 1959.

51 P. E. Hart, N. J. Nilsson, and B. Raphael, "A formal basis for the heuristic determination of minimum cost paths," *IEEE Transactions on Systems Science and Cybernetics*, vol. 4, no. 2, pp. 100–107, 1968.

52 E. Nikolova, M. Brand, and D. R. Karger, "Optimal route planning under uncertainty," in *6th International Conference on Automated Planning and Scheduling (ICAPS)*, vol. 6, pp. 131–141, 2006.

53 A. Orda and R. Rom, "Shortest-path and minimum-delay algorithms in networks with time-dependent edge-length," *Journal of the ACM (JACM)*, vol. 37, no. 3, pp. 607–625, 1990.

54 B. Ding, J. X. Yu, and L. Qin, "Finding time-dependent shortest paths over large graphs," in *11th International Conference on Extending Database Technology: Advances in Database Technology*, pp. 205–216, ACM, 2008.

55 M. P. Wellman, M. Ford, and K. Larson, "Path planning under time-dependent uncertainty," in *11th Conference on Uncertainty in Artificial Intelligence*, pp. 532–539, Morgan Kaufmann Publishers Inc., 1995.

56 W. Huang and J. Wang, "The shortest path problem on a time-dependent network with mixed uncertainty of randomness and fuzziness," *IEEE Transactions on Intelligent Transportation Systems*, vol. 17, pp. 3194–3204, 2016.

57 E. Miller-Hooks, "Adaptive least-expected time paths in stochastic, time-varying transportation and data networks," *Networks: An International Journal*, vol. 37, no. 1, pp. 35–52, 2001.

58 B. Y. Chen, W. H. Lam, Q. Li, A. Sumalee, and K. Yan, "Shortest path finding problem in stochastic time-dependent road networks with stochastic first-in-first-out property," *IEEE Transactions on Intelligent Transportation Systems*, vol. 14, no. 4, pp. 1907–1917, 2013.

59 A. A. Prakash, "Pruning algorithm for the least expected travel time path on stochastic and time-dependent networks," *Transportation Research Part B: Methodological*, vol. 108, pp. 127–147, 2018.

60 H. Frank, "Shortest paths in probabilistic graphs," *Operations Research*, vol. 17, no. 4, pp. 583–599, 1969.

61 X. Wu and Y. M. Nie, "Modeling heterogeneous risk-taking behavior in route choice: a stochastic dominance approach," *Procedia-Social and Behavioral Sciences*, vol. 17, pp. 382–404, 2011.

62 Z. Cao, H. Guo, J. Zhang, D. Niyato, and U. Fastenrath, "Finding the shortest path in stochastic vehicle routing: a cardinality minimization approach," *IEEE Transactions on Intelligent Transportation Systems*, vol. 17, pp. 1688–1702, 2016.

63 Z. Dai, X. Ma, and X. Chen, "Bus travel time modelling using GPS probe and smart card data: a probabilistic approach considering link travel time and station dwell time," *Journal of Intelligent Transportation Systems*, vol. 23, no. 2, pp. 175–190, 2019.

64 D. Aleksandrov and I. Penkov, "Energy consumption of mini UAV helicopters with different number of rotors," in *11th International Symposium" Topical Problems in the Field of Electrical and Power Engineering*, pp. 259–262, 2012.

Abbreviations

Full term	Abbreviation
Unmanned aerial vehicle	UAV
Field of view	FoV
Signal-to-noise-ratio	SNR
Quality of service	QoS
Quality of coverage	QoC
Line of sight	LoS
Art Gallery Problem	AGP
Ground node	GN
User equipment	UE
Wireless sensor network	WSN
Internet Service Provider	ISP
Internet of Things	IoT
Base station	BS
Road-side units	RSUs
Travelling salesman problem	TSP
Vehicle routing problem	VRP
Probability density function	PDF
Cumulative distribution function	CDF

Autonomous Navigation and Deployment of UAVs for Communication, Surveillance and Delivery, First Edition. Hailong Huang, Andrey V. Savkin, and Chao Huang.
© 2023 The Institute of Electrical and Electronics Engineers, Inc. Published 2023 by John Wiley & Sons, Inc.

Index

Autonomous Navigation and Deployment of UAVs for Communication, Surveillance and Delivery,
First Edition. Hailong Huang, Andrey V. Savkin, and Chao Huang.
© 2023 The Institute of Electrical and Electronics Engineers, Inc. Published 2023 by John Wiley & Sons, Inc.

Printed and bound by CPI Group (UK) Ltd, Croydon, CR0 4YY

16/04/2025

14658584-0002